Automated Stream Analysis for Process Control

VOLUME 2

CONTRIBUTORS

GARY L. BAKER

GEORGE F. ERK

JOHN C. HUDELSON

DAN P. MANKA

KARL J. SIEBERT

LEONARD J. WACHEL

AUTOMATED STREAM ANALYSIS FOR PROCESS CONTROL

VOLUME 2

EDITED BY

DAN P. MANKA

Pittsburgh, Pennsylvania

1984

ACADEMIC PRESS

(Harcourt Brace Jovanovich, Publishers)

Orlando   San Diego   New York   London
Toronto   Montreal   Sydney   Tokyo

7221 - 448X

ACADEMIC PRESS, INC.
Orlando, Florida 32887

United Kingdom Edition published by
ACADEMIC PRESS, INC. (LONDON) LTD.
24/28 Oval Road, London NW1 7DX

Library of Congress Cataloging in Publication Data

Main entry under title:

Automated stream analysis for process control.

Includes bibliographies and indexes.
1. Chemical process control--Automation. I. Manka,
Dan P.
TP155.75.A88 1982   660.2'81   82-8822
ISBN 0-12-469001-7 (v. 1)
ISBN 0-12-469002-5 (v. 2 : alk. paper)

PRINTED IN THE UNITED STATES OF AMERICA

84 85 86 87   9 8 7 6 5 4 3 2 1

Contents

13   Calibration Methods for Process Analyzers

Gary L. Baker

14   Interfacing Analyzers to Computers

John C. Hudelson

15   Process Analyzer Optimization in Control Systems

Leonard J. Wachel

16 Engineering of Total Analyzer Systems

George F. Erk

17 Process, On-Stream, and Chromatographic Measurements in Brewing

Karl J. Siebert

18 Reduction of H$_2$S and HCN in Coke Oven Gas

Dan P. Manka

19 Conversion of H$_2$S to Sulfur or Sulfuric Acid

Dan P. Manka

20 Analysis of Chemicals Derived from Coal Carbonization

Dan P. Manka

List of Contributors

Numbers in parentheses indicate the pages on which the authors' contributions begin.

GARY L. BAKER, (1), Process Measurement Systems, Engineering and Technology Services Division, Union Carbide Corporation, South Charleston, West Virginia 25303

GEORGE F. ERK, (113), Sun Refining and Marketing, Philadelphia, Pennsylvania 19103

JOHN C. HUDELSON, (63), Amoco Oil Company, Whiting, Indiana 46394

DAN P. MANKA, (169, 183, 187), Pittsburgh, Pennsylvania 15218

KARL J. SIEBERT, (139), The Stroh Brewery Company, Detroit, Michigan 48226

LEONARD J. WACHEL, (93), UIC, Incorporated, Joliet, Illinois 60434

Preface

If a chemist or chemical engineer is authorized to develop a process-control method, procedures that are discussed in Volume 1 of "Automated Stream Analysis for Process Control" may be adapted, perhaps with some modification. These procedures are the result of many tests by the authors, who have successfully eliminated the many negative results normally associated with the development of a new process-control method. Although these successful methods may not directly apply to the process that the chemist or chemical engineer is examining, it is shown that there are several approaches to choosing the sample point, sample preparation, sample transport, and analyzer best suited for the components in a specific process stream.

This second volume describes the engineering approach to the design of a process-control system and the interfacing of the analytical results with computers or printouts to apprise the operator of the progress of the stream operation.

It is well known that the accuracy of any analytical instrument is no better than the calibration of that instrument with proper standards. The discussions and procedures outlined for calibration in this book are the most comprehensive and thorough found in any literature on the subject. The chemist or chemical engineer will find many calibration methods together with their advantages and disadvantages. This discussion should also be of interest to those who have calibrated instruments in the past. They could learn some new techniques that may enhance the accuracy of their calibration procedure.

The book has an in-depth discussion of the chemical reactions and scope of analytical procedures used in the brewing of a popular beer.

This discussion illustrates that practically every process can be made more profitable by implementing continuous analytical procedures to monitor the various reactions occurring in the process.

In the environmental field, the reduction of hydrogen sulfide and hydrogen cyanide in fuel gas before the oxidation of sulfur compounds to sulfur dioxide during combustion is an important approach to the reduction of acid rain now believed to be destroying lakes and land in the Eastern United States and Canada.

Contents of Volume 1

13

Calibration Methods for Process Analyzers

GARY L. BAKER

Process Measurement Systems
Union Carbide Corporation
South Charleston, West Virginia

I. Introduction

A. *Importance of Calibration*

Process analyzers are valuable tools in the operation of production units even if they are able to supply only trend data. Increasingly, however, automatic process analyzers are used to provide accurate, quantitative process-stream data directly to controllers or sophisticated control systems. For these applications, the analyzers should have good precision and accuracy and operate with minimum maintenance for periods of many months or years.

The user can ordinarily determine whether an analyzer is accurate when received only by checking it with a series of calibration standards. After this initial calibration, the analyzer must be periodically checked with at least one suitable standard to assure continued accuracy.

B. *Scope of Chapter*

Most of the calibration methods and types of calibration standards discussed in this chapter have been applied by the author for calibrating analyzers such as gas chromatographs (GCs). For simplicity, most of the discussion in this chapter is directed toward applying these methods to process GCs. However, many of the methods are not limited to GCs and are described for virtually any conceivable type of liquid or gaseous stream and for nearly any type of composition-measuring analyzer. Primary emphasis is placed on practical calibration methods and standards for process analyzers.

II. Calibration Methods

A. General Considerations

1. SAFETY

Most analyzers are supplied with instruction manuals that describe hazards associated with the operation or maintenance of the analyzer. Sometimes, however, information pertaining to the safety of calibration is not included. The user should request this information from a reliable source.

The user should become familiar with the hazardous properties of any of the chemicals which are contained in any standards used to calibrate an analyzer. Caution should be used when preparing samples to avoid contacting toxic substances and mixing chemicals that might react violently. Caution should also be used in handling and using gases contained at high pressures. The user should make sure that pressure regulators and other components are suitable for the types of materials and pressures to which they will be exposed. Relief valves should be used if there is any danger of overpressurizing these components.

2. WEIGHT VERSUS VOLUME BASIS

Confusion often arises concerning whether analytical results are provided on a weight or volume basis. These results are normally desired on the weight basis (i.e., weight percentage) because this is more convenient for the determination of material balances. Unfortunately, virtually all process analyzers can directly provide results only on a molar or volume basis, because a fixed volume of sample is actually analyzed. If the sample density is constant, then the basis is not important because constant volume would equate to constant weight.

Sample density is not always constant; therefore, errors can result. The following example illustrates what can happen: Assume a sample size of 0.001 mole (22.4 ml) at standard temperature and pressure (0°C and 760 Torr). Now assume a sample containing 10% nitrogen by volume in methyl chloride. This sample would contain 2.8 mg of nitrogen and 45.5 mg of methyl chloride, or 5.8% nitrogen by weight. Now assume a sample with 10% nitrogen and 10% helium (both by volume) in methyl chloride. This sample would also contain 2.8 mg of nitrogen, but it would be 6.4% of the sample weight. Addition of the helium would cause a change in the weight percentage of nitrogen in the sample from 5.8 to 6.4% (a 10.3% relative error), while the volume percentage would remain constant.

There is a way out of the dilemma. The normalization method (Section II,C) permits the results to be calculated on the weight basis. Most modern process GCs and some other types of analyzers can use the normalization method, but all of the significant compounds in the samples must be measured to apply it.

The internal standard method (Section II,D) also permits the results to be calculated on the weight basis. Unfortunately, this method is more difficult to apply to process analyzers.

3. PEAK AREA VERSUS PEAK HEIGHT

Some analyzers, especially GCs, measure each component as a separate, Gaussian-shaped peak. The concentration of each component is approximately linearly proportional to its peak area for most GC detectors, if the column and the detector are not overloaded (exposed to too much sample). The peak areas are not significantly affected by measurement parameters, such as column flow rate or temperature, although they may be affected by *detector* parameters such as temperature or flow rate. The relative area responses of various compounds are essentially unaffected by any measurement parameters. Therefore, the use of peak areas is very useful for the calibration of GCs (Kipiniak, 1981).

Concentration is also approximately linearly proportional to the peak height (again within restricted operating ranges) if all measurement parameters are closely controlled. The peak height, however, is influenced considerably by column parameters, especially temperature. This greatly limits the applicability of the normalization method of calibration.

Thus the peak-area method of measurement is often preferred to the peak-height method because it permits more calibration-method options. However, it requires the use of an integrator, which may not always be available. Also, better repeatability can sometimes be obtained by the peak-height method, especially if the peak-detection parameters are not highly optimized for the integrator used for the area measurements. The choice of peak area or peak height must be made for each application.

4. TYPICAL ANALYZER CALIBRATION

Normally, the determination of the accuracy of an analytical measurement system requires the introduction of one or more calibration standards. A suitable method for accomplishing this in the field should be considered before an analyzer is purchased and installed (see Section V,A).

Analyzers that do not require a high degree of accuracy or have been determined to be acceptably linear may require only the adjustment of zero and span analyzer controls. (Refer to the instruction manuals supplied with each analyzer for specific calibration instructions.) For example, an infrared spectrometer can be calibrated by introducing a zero standard that does not contain any of the component being measured and adjusting the zero control to obtain a zero reading from the analyzer's output indicator. Then a calibration standard containing a known concentration of the desired component is introduced, and the span control is adjusted to obtain an output from the analyzer corresponding to the sample concentration.

Sometimes, a good zero standard is not available (i.e., it contains some of the component being measured). When this occurs, it is necessary to adjust the zero control so that the analyzer indicates the concentration of the measured component in the zero standard. Then it is likely that the zero and span analyzer controls will interact. (These controls may interact on some analyzers regardless of the quality of the zero standard.) Therefore, it is recommended that the analyzer zero and span be rechecked after any adjustments are made to ensure that such an interaction does not introduce an error.

After this procedure has been completed, the analyzer will produce an analog or digital signal that will be indicative of the desired component concentration.

The following example typifies the procedure: Assume that an analyzer is to be calibrated for 0 to 10% ethanol in water. The analyzer's zero control should be adjusted to produce an analyzer response of zero when a sample of pure water is introduced. Then a calibration standard containing 10% ethanol in water should be introduced, and the analyzer span should be adjusted so that the analyzer's output indicator reads full scale (10% ethanol). The zero and span standards should be rechecked until no further adjustments are necessary. Thereafter, when a sample containing an unknown concentration of ethanol in water is analyzed, the ethanol percentage

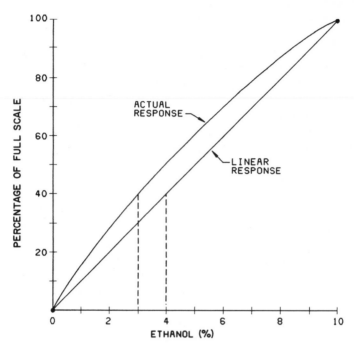

Fig. 1. Actual analyzer response versus linear response.

can be read directly from the indicator. For example, 4% ethanol would be indicated as 40% of the full-scale value of the indicator.

In this example, if the response of the analyzer to ethanol is not linear, accurate readings will be obtained only at concentrations near those used to establish calibration. Figure 1 illustrates a case where a reading of 40% of full scale would represent only 3% ethanol instead of 4%, resulting in a relative error of 25% if linear response is assumed. This kind of error may be acceptable for some applications but totally unreasonable for others (see Section II,A,6).

The initial calibration of an analyzer that must produce better accuracy will require the use of several calibration standards covering the concentration range of interest. These standards are required to characterize the response of the analyzer to various concentrations of the component(s) to be measured. This can best be done by regression analysis (Gordon and Ford, 1972; Draper and Smith, 1966). Also, standards should be used to determine the influence of other compounds in the sample on the component(s) measured. Sometimes a nonmeasured component can cause errors in a measured component. For example, when two components pass through a nonspecific detector at the same time, the analyzer may not distinguish between them.

Recalibration of an analyzer in the field will ordinarily require only one or two standards (zero and span) unless the analyzer response has changed significantly since initial calibration. However, if significant response changes have occurred (e.g., over 20%), then one or more additional standards should probably be used to verify that the initial characterization of response is still valid.

The types of standards selected for calibration will depend on the accuracy requirements. The National Bureau of Standard's Standard Reference Materials (NBS's SRMs) are probably the best available (see Section III,A). Ordinarily, certified standards should be acceptable.

Sometimes an analyzer may be accurately calibrated but still produce poor results. The sampling system used to deliver the process sample to the analyzer can be responsible for such a problem (see Section VI,B). Comparison of the response of a calibration standard when introduced directly to the analyzer with the response after the standard has flowed through the sampling system will usually reveal if such a problem exists on a particular system.

5. FREQUENCY OF CALIBRATION

One must consider the possibility of changes in the analyzer's response characteristics over time because of the effect of ambient temperature or pressure changes, deterioration or malfunction of components within the analyzer, or other factors. Even the most stable analyzers will require occasional calibration checks.

The need for calibration may vary from more than once an hour to once a year; or calibration may be required whenever any maintenance is performed. The frequency should depend on how important accurate calibration is for the particular application.

For those applications where it is determined that calibration should be checked frequently (even once a month could be considered frequently), then some means for calibration should be provided with the analyzer. In some cases, calibration hardware should be built into the analyzer.

6. LINEARIZATION

Whenever a high degree of accuracy is required, it is necessary to characterize the response of the analyzer to each compound measured at various concentrations from the minimum to the maximum expected in the process stream. This is best done by a regression analysis of the data (Gordon and Ford, 1972; Draper and Smith, 1966).

Ordinarily, the response will be linear over the range of interest. If it is not linear, the response will often fit the quadratic equation

$$C = aX^2 + bX + c, \tag{1}$$

where C is the concentration of the compound measured, X the value of the detector response, and a, b, and c are constants empirically derived from the experimental data.

Most process analyzers have the capability of indicating true concentration by using Eq. (1) internally. Thus they provide accurate, linearized output. After initial calibration, these analyzers can ordinarily be recalibrated with acceptable accuracy by introducing a calibration standard and adjusting the b constant to provide the proper analyzer output. If the analyzer response has changed significantly since initial calibration, the adjustment of the b constant may not provide accurate calibration over wide concentration ranges. Complete recalibration with a set of calibration standards may be necessary.

B. Area-Percent Method

The area-percent method is the simplest calibration method for analyzers that have the capability; it can be utilized only with analyzers that measure the areas of all major sample constituents. It is most commonly utilized with GCs, since they often measure all (or virtually all) of the compounds in a sample, and it is the default method for most modern GCs.

The area percentage of a component is determined simply by dividing its peak area (typically in volts-seconds units) by the total area of all peaks resulting from a sample analysis. The area percentage of a compound can be considered as only an approximate indication of the actual weight or volume percentage of that compound unless the detector is equally responsive to all compounds. This is not normally the case, and errors of over 100% can result. It is normally useful for only very rough approximations of sample concentrations. It is also useful in the development of other calibration methods.

The major advantages of the method are that it does not require calibration standards and that sample size is not important within the operating range of the instrument.

C. Normalization Method

The normalization method of calibration (Keulemans, 1959) is very similar to the area-percent method. (Refer to Table I for a comparison of various common methods.) The only difference is that each peak area is corrected for its detector response, but this difference makes the normalization method much more useful. Peak heights can be used instead of areas if appropriate response factors are available (but see Section II,A,3).

Like the area-percent method, the normalization method does not require the introduction of calibration standards to correct for normal variations in detector response. Small variations in detector operating parameters have approximately the same relative effect on each compound. Also, variations in sample size, which can result from changes in sample pressure (when analyzing gas samples) or partial plugging of a sample port in a sample-inject valve, do not have a significant effect on accuracy if the detector response for each component is essentially linear. Accurate results can be obtained with nonlinear detector responses only if a linearization technique is applied to each peak area (see Section II,A,6).

Another advantage of the normalization method is that the results can be determined on a weight or volume basis with equal accuracy and ease by the equation

$$\text{percentage(Y)} = \frac{100 A_Y F_Y}{\sum_i A_i F_i}, \tag{2}$$

where Y is the component of interest, A_Y the area of the Y peak, F_Y the response factor for Y, and $\sum_i A_i F_i$ the summation of the product of area and response factor for all i components in the mixture. The percentage(Y) determined from Eq. (2) will be either on a weight or a volume basis, depending on whether the response factors are based on weight or volume.

Response factors can be calculated by the equation

$$F_Y = Q_Y / A_Y, \tag{3}$$

where Q_Y is the quantity of Y in weight or volume. Peak height can be used instead of area in Eqs. (2) and (3), if preferred. Also, relative response factors are more commonly used than absolute response factors; they are relatively unaffected by detector parameters. Relative response factors can be obtained by dividing the absolute response of each component of interest by the absolute response of an internal reference component (e.g., hexane). The same units must be used for all components.

TABLE I

SUMMARY OF CALIBRATION METHODS

Method	Application	Advantages	Disadvantages
Area percent	Normally used to aid in development of more-accurate methods for chromatographs; sometimes directly usable where great accuracy not required; can be applied to any analyzer where all components are measured	Easy to implement; does not require external standard; sample size is relatively unimportant	All components must be measured; integrator is required; not a true measure of composition
Normalization	Normally used for gas chromatographs; can be applied to any analyzer where all components are measured	Easy to use; does not require external standard; sample size is relatively unimportant; results can be reported on either weight or volume basis	All components must be measured; integrator is required
Internal standard	Normally limited to laboratory-type analyzers (ones that can measure two or more components on each sample injection).	Does not require external standard; sample size is relatively unimportant; results can be reported on either weight or volume basis; all components in sample do not have to be measured; reliability of results is maximized for each sample	Requires manual alteration of each sample

TABLE I (*Continued*)

Method	Application	Advantages	Disadvantages
Deferred standard	Primarily for process chromatographs	Reliability of results is maximized for each sample analyzed	Requires additional analyzer hardware; measurement parameters are more difficult to establish; consumption of calibration standard may be excessive; may not provide accurate results on weight basis
External standard	Universal	Can use with virtually any analyzer; only the components of interest must be measured	Sample size must be repeatable for accurate results; may not provide accurate results on weight basis; frequent recalibration may be required

Reasonable accuracy can often be obtained with the normalization method by using relative response factors reported in the literature (Dietz, 1967), being sure to check whether the reported factors should be used as multipliers or divisors. Factors for all the compounds of interest may not be available. Also, there is always some risk that the relative values may be slightly different for a particular detector (Roy *et al.*, 1980). Even the same detector, when its operating parameters are changed drastically, may yield different relative-response values (Crabtree and Blum, 1981).

Hardware problems can also cause errors. For example, if the sample loop in a sample inject valve preferentially absorbs one or more compounds, then significant errors can result. Also, if a particular compound reacts with the chromatographic column used in the analyzer, then lower results would be reported for that compound. For these reasons and others, if maximum accuracy is desired, then response factors should be determined by using the actual system.

The response factor for each component can be determined experimentally by establishing the response it generates in the pure state. A better way is to use a calibration standard containing the component in the approximate concentration at which it will be measured.

Sometimes, accomplishing this is not as simple as it might seem. It may not be possible to obtain some of the compounds in a reasonably pure state or with known composition. Also, in some cases it may be desired to analyze a sample stream in a vapor phase, but the sample must be heated to prevent it from condensing; this complicates the preparation and introduction of such samples. Conversely, it may be desired to analyze a liquid stream containing components that may vaporize unless the sample is kept under pressure. Methods for preparing and feeding such samples are discussed in Sections III and IV.

Sometimes it is not possible or practical to analyze all the significant compounds in a sample stream with one analyzer. In such cases, the normalization method cannot be used unless two or more analyzers are used, and then additional instrumentation may be required to perform the computations. The additional cost and complexity would probably outweigh any advantages.

When the normalization method is used, the analyzer should be occasionally checked with an accurate calibration standard to ensure that accurate data is obtained.

D. Internal-Standard Method

The internal-standard (IS) method (Nowicki *et al.*, 1979) requires that a precisely determined quantity of an extra compound, one not present in the original sample, be added to each sample analyzed. This is rather easily accomplished with laboratory-type analyzers but normally impractical with process-type analyzers. It will be discussed here because laboratory-type analyzers, especially those with automatic carousel samplers, are sometimes used within process units.

Like the normalization method, the IS method requires the use of response factors. However, instead of comparing the response of each measured compound with the total response, it ratios the response of each compound with the response of the IS. The actual equation used to calculate the weight percentage of a compound Y is

$$\text{Wt. percentage(Y)} = \frac{100 A_Y F_Y Q_s}{A_s F_s Q_t}, \qquad (4)$$

where A_Y is the area of peak Y, A_s the area of the IS, F_Y the response factor for compound Y, F_s the response factor for the IS,

Q_s the weight of the IS added to the sample, and Q_t the total weight of the sample excluding Q_s. Peak heights can be used instead of areas if appropriate response factors are used (but see Section II,A,3).

This method can be very accurate for liquid samples. Also, it does not require the measurement of the compounds; only the compounds of interest and the IS must be measured. More than one IS may be used if desired (e.g., one for each compound of interest). Results can be reported on a volume-percentage basis if preferred.

The IS must be well separated from all other compounds in each sample to obtain good accuracy. Accuracy is unaffected by sample size if detector response to the compounds measured is essentially linear. As with the other calibration methods, it is desirable to sample a calibration standard periodically to verify proper functioning of the entire analyzer.

E. *Deferred-Standard Method*

The deferred-standard (DS) method (Guillemin, 1980) is very similar to the IS method except that it is much more suitable for process GCs. By the DS method, a known quantity of a pure compound (the DS) is injected during each analysis cycle. The injection of the DS occurs at a different time than the process-sample injection so that it does not interfere with any component in that sample. It can be injected by the sample inject valve used for the process sample or by an alternative valve that is in series with the process-sample inject valve.

The DS method requires the use of response factors. The molar, or volume, concentration of any component in the sample can be determined by using Eq. (4). Weight percentage cannot be determined directly because fixed volumes of sample and the DS are injected (see Section II,A,2). Therefore, Q_s and Q_t should be on the volume basis. Also, Q_t does not include the volume of the DS.

F. *External-Standard Method*

All the previously discussed methods do not require the sample size to be exactly constant and are somewhat tolerant of changes in analyzer conditions such as column temperature (in a GC). With these methods, responses to different compounds are compared within each sample, providing compensation for such effects. The external-standard (ES) method is different; it requires that the response to one or more compounds in each sample analyzed be compared with responses to a previously analyzed calibration standard. For example, if the size of a particular sample is 10% smaller, the indicated concentration for each compound measured will be approximately 10% lower.

Most modern GCs are capable of operating for long periods of time with little variation in sample size or detector response. This means that the ES method can be used with accurate results. In fact, it is a very common calibration method for process GCs. Also, it is applicable to virtually all types of analytical instruments, such as spectrometers, where most other methods are not.

The calculations required for the ES method are very simple. Once the response factor for a compound is determined, its concentration in an unknown sample is determined by simply multiplying the factor by the area (or peak height) response for that compound.

A major advantage of the method is that only the compounds of interest need be calibrated and measured. The major disadvantages are that the analyzer must perform precisely or repeatably and that frequent recalibration (possibly once a day or even once an hour) may be necessary to obtain accurate results. Also, results can be determined on a volume basis only unless sample density is constant (see Section II,A,2). This is frequently ignored by analyzer vendors and many users.

This method can be used to great advantage in conjunction with the normalization method for some analyzers (see Section II,C).

G. *Miscellaneous Methods*

1. ELECTROLYTIC CELL

Some analyzers can be calibrated by making use of well-known physical laws. For example, an electrolytic cell is sometimes used for the analysis of water in a gas stream. In an electrolytic cell, water is dissociated into hydrogen and oxygen, and a corresponding current is measured as defined by Faraday's law. This law states that one gram-equivalent weight of matter is altered at each electrode for one Faraday of electricity passing through the electrolyte (Weast *et al.,* 1964). The mass of water electrolyzed per unit of time is then directly proportional to the current. If the sample flow rate through the cell is held constant, the volume percentage of water can be easily calculated.

There is usually a catch to these ideal-sounding measurement and calibration methods, and the electrolytic cell is no exception. Accurate results will be obtained only if all the water in the sample is electrolyzed, and this is sometimes difficult to achieve. Therefore, it is desirable to check the calibration of such an instrument with some other method, such as the external-standard method.

2. DEW POINT

Another essentially self-calibrating measurement is the dew point (DP) method (Cole and Reger, 1969). By this method, the gas sample is cooled until dew begins to form on a polished surface. At this DP temperature, the vapor pressure of water can be readily determined from available physical-property data. This pressure is the same as the partial pressure of the water in the gas mixture. By Dalton's law, which is valid under only ideal gas conditions, the molar concentration of water is the ratio, multiplied by 100, of the partial pressure to the total pressure at which the DP is determined. For example, at a DP of 7.1°C, the partial pressure of the water is 7.6 Torr. At a system pressure of 760 Torr, the percentage water would be $(7.6/760) \times 100 = 1.0\%$.

3. GAS-DENSITY BALANCE

A third self-calibrating method involves the use of a gas-density balance (Guillemin, 1980). This device is actually a GC detector having a response that is directly proportional to the difference in density of the component it is detecting and the carrier-gas density. Therefore, once its response is determined for one component, its response to any other component can easily be calculated from density data alone. This detector has great potential for use as a calibration aid.

4. STANDARD ADDITION

The standard-addition method (Williard *et al.*, 1974) is a common method used in the laboratory to establish the concentration of one or more minor components in a matrix. This method can be applied to certain analyzers such as GCs for determining the composition of a mixture that is to be used as a calibration standard.

5. LABORATORY ANALYSIS

A common method for field calibration is to pull samples for lab analysis and note the analyzer reading at the time the sample is taken. Then, after laboratory results are received, the analyzer sensitivity is adjusted accordingly. This method may work well for a very slowly changing process but is impractical where process conditions vary widely and rapidly.

6. OTHER METHODS

A common method for field calibration of spectrometers or other optical analyzers involves the use of neutral-density filters. This method is not suitable for initial calibration but is useful for determining if any optical or electronic drift has occurred since initial calibration. Calibration checks are performed by inserting a neutral-density filter into the optical path to simulate the absorbance by a sample of known composition.

Other calibration methods include gas- and liquid-phase titration. These methods would normally be adequately described by the manufacturer of instruments that can conveniently be calibrated by these special methods; therefore they will not be described in this chapter.

H. *Redundant Methods of Calibration*

The experience of the author and others, such as Stedman *et al.,* (1976), indicates that one should not rely on a single method of calibrating an analyzer if accuracy is very important. The author has experienced several problems over the years that would have resulted in considerable error if one or more redundant methods had not been used to verify the results of the principal method. For example, sometimes gas calibration standards do not contain even approximately the concentration of one or more components indicated. On one occasion, two certified standards, which were ordered at the same time, had their labels switched. The use of redundant methods enabled us to recognize the error before serious problems resulted.

A safe rule to follow is to suspect problems with any calibration method or standard until it is compared with another completely independent method or standard. For example, if a commercial calibration standard is used to calibrate a GC, then determine the area response factors for each component by using the standard, and compare these factors with response factors published in the literature (Dietz, 1967). If the relative response factors compare closely, then there is good evidence that the commercial standard is accurate.

III. Calibration-Standard Preparation

A. *Definitions of Standards and Mixtures*

There is not universal agreement on the definitions of primary and secondary standards pertaining to calibration of analytical instruments. The purist would define a "primary standard" as a standard unit of a measurable quantity, such as length or mass, which is maintained at a national laboratory or institution. This "primary standard" is itself an arbitrary quantity. However, the U.S. NBS maintains such Primary Standards that are accepted throughout the world.

None of these Primary Standards can be used directly to calibrate composition-measuring analyzers. The best-possible calibration method or standard is only traceable to these Primary Standards and therefore, strictly speaking, would be a secondary standard. The NBS provides SRMs that are directly traceable to these Primary Standards.

Another definition of "primary standard" is "the top standard in any measurement system." By this definition, the NBS's SRMs would qualify as primary calibration standards (Hughes, 1976). Also, any method defined by the NBS as directly traceable to a true NBS Primary Standard might qualify as a primary standard.

Many suppliers of gas standards have other definitions. They use the term "primary standards" for gas mixtures that are prepared gravimetrically by weighing the quantity of each gas added using NBS certified weights. They use the term "certified standards" for mixtures prepared by any desired technique and then analyzed by analyzers having been calibrated with gases that they consider "primary standards." These certified standards are actually secondary standards by their definition.

The term "mixture" will be used to describe a sample that does not qualify as a primary or secondary standard.

It is the author's opinion that certified standards are adequate for the calibration of most process analyzers. If the utmost in accuracy is required, then SRMs should be used. Where SRMs are not available, gravimetrically prepared standards should be used, although they should be analyzed to maximize reliability.

B. Static Gas Mixtures

Two types of static gas mixtures are commonly used for calibration: (i) those prepared at atmospheric pressure and (ii) those prepared at pressures above atmospheric pressure (Roccanova, 1968).

1. NONPRESSURIZED MIXTURES (VOLUMETRIC DILUTION)

Calibration standards prepared at atmospheric pressure (Fig. 2) are ordinarily prepared in a gas bag or large syringe made of relatively nonabsorptive materials. Standards of this type are prepared by the user, and they are convenient for analyzers that pull their own samples. They are not available commercially for several reasons: the quantity of gas in a conveniently sized container is small (e.g., less than 10 liters); the cost per unit volume of gas is very high; and the composition of the mixtures usually changes substantially within a short period of time (e.g., 1 day) because of absorption into the walls of the container or permeation through the walls. The materials commonly used in bag construction are Tedlar or Mylar, sometimes with an aluminum coating to minimize permeation. The syringes are most commonly constructed of a plastic material such as Plexiglas.

The containers are ordinarily filled by adding known volumes of the desired gases; hence, this method is sometimes called the volumetric-dilution method. The desired volume of each gas can be added by syringe or by any desired metering method. Liquids can even be added by syringe if they have sufficient vapor pressure to be in the vapor state in the final mixture. The required quantity of liquid can easily be calculated after the volume of the vapor required in the final mixture is determined, assuming an ideal-gas mixture, using the equation

$$V_L = V_g(MW)/22,400D, \tag{5}$$

where V_L is the volume of liquid in milliliters, V_g the volume of vapor in milliliters required at standard conditions of 0°C and 760 Torr, MW the molecular weight of the compound, and D the density

Fig. 2. Volumetric-dilution chamber. [Reproduced from Baker and Brubaker (1974) by permission of Instrument Society of America.]

of the liquid in grams per milliliter for the temperature at which V_L is measured.

The volume of each gas required to obtain a desired concentration in the final mixture is determined by the relationship

$$V_g = V_t \frac{C}{100} \frac{273.2}{T} \frac{P}{760},$$ (6)

where V_t is the total volume of gas mixture in milliliters at T K and P Torr; C the desired concentration of gas in volume percentage; T the temperature of the gas mixture in Kelvins; and P the pressure of the mixture in Torr. Equation (6) is valid at low pressures, where the ideal gas law is applicable. Temperature and pressure corrections can be ignored if all gases are added at the same conditions.

The volumetric dilution method is not highly accurate, although accuracies within 5% of the target concentrations can be obtained with care. The main value of the method is that it is easy to implement even under field conditions and can be used to indicate if another supposedly more accurate method is working properly.

2. PRESSURIZED MIXTURES

The pressurized static-sample method is probably the most widely used method of calibrating analyzers with gas standards. The reasons are that these standards, once prepared, are very convenient to use. They are normally stable (the gas composition does not change significantly with time), relatively inexpensive, and available with accuracies within 1% of the stated concentration of each component. A large variety of gas mixtures is available from several specialty gas suppliers.

A large number of factors can affect the accuracy of this type of standard. Expensive equipment and technicians with a great deal of skill and experience are required to prepare them with a high degree of accuracy. In addition, some gas mixtures are dangerous to prepare, although the final mixture might be perfectly safe to use (see Section II,A,1). Therefore, it is not recommended that most users attempt to prepare this type of sample. Information is provided in this chapter primarily so the user can obtain an idea of many of the factors affecting the accuracy of such samples, can more intelligently specify such samples to meet specific needs, and can be aware of problems that can occur with certain mixtures.

a. Partial pressure method. Many pressurized gas standards are prepared by the partial-pressure method whereby the composition of the gas mixture is assumed to be directly proportional to the partial pressure of each component. The concentration of each gas can be determined by the equation

$$C = 100P_c/P_t, \tag{7}$$

where C is the concentration of the gas of interest, P_c the partial pressure of the gas of interest, and P_t the total pressure of the gas mixture.

Preparation of the gas mixture is facilitated by the use of precise absolute-pressure gauges.

An example of how one would prepare a simple mixture of 1% acetylene in ethylene at a total pressure of 100 psia is as follows:

(i)   evacuate the container to be used for the gas mixture;
(ii) add acetylene until the cylinder pressure is 1 psia; and
(iii) add ethylene until the cylinder pressure is 100 psia.

Although this mixture is very simple and would be easy for an experienced technician with suitable equipment to prepare, a novice could end up with a mixture that is very inaccurate. Factors that influence accuracy include precision of the pressure gauges, gas purity, temperature of the gas cylinder, contamination of the gas-mixing equipment (hoses, etc.), the condition of the interior surface of the gas cylinder, incomplete evacuation of the cylinder, gas compressibilities, and mixing technique. These factors become increasingly important as the complexity of the gas mixture increases; factors such as more components in the mixture, lower concentrations of some components, reactivity, low volatility, and higher final pressure could tend to magnify any error.

Typical inaccuracies that can result from some of these factors are described below, using the example of the acetylene in ethylene mixture.

A 0–30-psia precision pressure gauge would permit an error of up to 0.03 psi in the pressure of acetylene added. A 0–300-psia precision pressure gauge would permit an error of as much as 0.3 psi in the ethylene pressure. These could result in an error of up to 3% of the quantity of acetylene and 0.3% of the quantity of ethylene added. The total compounded error could be as great as 3.31% of the actual quantity of acetylene. This error could be greatly reduced by a more precise measurement of the acetylene pressure with a device such as a servomanometer.

A common technique for reducing the errors associated with limitations in the resolution and precision of the pressure-measuring equipment is the use of base stocks. A base stock is actually a gas standard containing one or more of the lower-concentration components and the background gas (or matrix) that are to be contained in a final gas standard. For example, using the acetylene in ethylene mixture described above, a base stock containing 10% acetylene in ethylene could be prepared first. Some of it could then be used in the preparation of the final mixture. The use of a base stock permits the addition of small quantities of the desired component by adding a larger quantity of the gas mixture; thus errors caused by poor resolution and poor precision of the pressure-measuring equipment are less significant.

The acetylene used to prepare the standard may contain as much as 3% acetone since it is shipped as a solution in acetone. This could cause an additional 3% relative error in the quantity of acetylene added unless the acetone is removed or accounted for when preparing the standard.

A cylinder temperature change of 1°C (at 20°C) would cause a relative error of approximately 0.34% of the actual quantity of acetylene. The temperature of the gas added to a cylinder ordinarily increases because of the release of energy when it is compressed. The temperature increase can easily exceed 10°C if the gas is added rapidly to a high pressure.

The nonideal compressibility of ethylene at 100 psia would cause a relative error of approximately 2.5% of the quantity of ethylene, unless accounted for when preparing the sample. This would also result in a relative error of up to 2.5% of the quantity of acetylene in the mixture.

Mixtures which contain one or more compounds with low vapor pressure can be very difficult to prepare. The desired compound can usually be added as a vapor. However, if its partial pressure is equal to or greater than its vapor pressure, condensation can result. Also, when the next gas is added, it might not mix thoroughly. If this occurs, the layer of fresh gas might act like a piston pushing against the first gas, increasing its partial pressure above its vapor pressure in a portion of the cylinder. This would cause some of the first compound to condense; this would, in turn, reduce the indicated cylinder pressure, possibly causing a considerable error in the gas composition. Also, considerable time may be required to revaporize the compound. If the mixture is used before it is totally vaporized and mixed with the other gases, even greater errors could result.

Even if a mixture with condensable gases has been accurately prepared, the user could have problems if the ambient temperature is too low. Some of the less volatile compounds could condense, and any gas used from the cylinder would not be representative of the composition as prepared.

b. Gravimetric method. The gravimetric method of preparing gas calibration standards is now widely used by specialty gas suppliers. This method requires accurate weighings of the cylinder used to contain the mixture before and after each gas is added. The number of moles of each gas can then be calculated by dividing the weight of each gas by its molecular weight. The molar, or volume, percentage composition can then be easily determined by the equation

$$C_X = 100 M_X / M_t, \tag{8}$$

where C_X is the concentration of component X, M_X the number of moles of component X, and M_t the total number of moles in the mixture.

Although the actual composition of the gas is determined from the weights of each gas added, the preparation of the sample is sometimes facilitated by observing the pressure increase as each component is added. Accurately weighing the sample and cylinder while adding gases is difficult because the hose used to add the gas can contribute substantially to the cylinder weight.

Maximum accuracy by the partial-pressure method requires the final sample pressure to be kept low to reduce compressibility errors. Compressibility of the gases has no effect on the accuracy of the gravimetric method. Also, the absolute precision and accuracy of the balances used to determine the weights are not affected appreciably as the weight is increased. The quantity of each gas added is greater at higher pressures; therefore, the accuracy of a mixture prepared by the gravimetric method is better at higher pressures.

Balances are available that have the capability of weighing a load of up to 75 kg, which is sufficient for most gas standards, with accuracy and readability within 1 mg. Typically, over 5 kg of gas are added to a sample. Therefore, it would appear that one could determine the total quantity of the gas within 0.2 ppm. However, the uncertainty of weighing a gas in a large cylinder may be closer to 1 g than 1 mg, primarily because of buoyancy effects of air on determining the precise weights.

The cylinder is partially buoyed by the sea of air around it. There-fore, if the density of the air changes (when using a single-arm balance), the weight of the cylinder in air changes. Changes in baro-metric pressure, air temperature, and humidity all affect air density. For example, assuming a cylinder with a volume of 50 liters, a change in barometric pressure of 1 Torr would cause the apparent weight to change by approximately 80 mg. A change in air tempera-ture of 1°C would change the apparent weight by approximately 220 mg. A change in the dew point of the air from 10 to 11°C would change the apparent weight by approximately 20 mg.

The use of a double-arm balance would theoretically eliminate these errors if a cylinder of precisely the same volume were used on the second arm of the balance. However, unless the conditions were exactly the same for both cylinders, errors would still result. In practice, it is very difficult to maintain a uniform temperature distri-bution within a space large enough to contain even one cylinder. Adding a gas normally increases the cylinder temperature. There-fore, temperature deviations of as much as 1°C might occur unless the air is thoroughly mixed by using fans or blowers. Unfortunately, turbulent air around the balance also can introduce significant errors in weighing. Therefore, the double-arm balance may actually pro-vide little advantage in improving accuracy.

For many samples, it is necessary to minimize the total cylinder pressure because of the low vapor pressure of one or more of the components in the sample. Also, some components may represent only a small part of the total composition of the gas. In these cases, only a few grams (even less than 1 g) of one particular component may be added. If, for example, only 2 g of a component are added and the uncertainty of weighing the sample is 0.2 g, then a 10% relative error in the quantity of that component could result. The use of a base stock (as discussed in Section III,B,2,a) would then be neces-sary to obtain improved accuracy of that component.

Other factors that may affect the accuracy of a sample prepared by the gravimetric method are similar to those affecting partial-pressure samples. These factors include impurities in the gases, absorption by the cylinder walls, and the technique used for prepar-ing the sample.

3. STABILITY OF STATIC GAS MIXTURES

Gas mixtures of reactive or highly absorptive gases, such as sulfur compounds, oxides of nitrogen, carbon monoxide, and water vapor, tend to be unreliable when contained in cylinders; researchers have documented problems associated with both accuracy and stability of such mixtures (Lee and Paine, 1976; Wechter, 1976). The concentrations of these components in the gas delivered from a cylinder can vary with time, temperature, or pressure. Unfortunately, the nature of the instability of such mixtures is undefined. One cylinder may deliver a relatively stable gas mixture, but another supposedly identical cylinder may deliver a totally unsuitable mixture. Some suppliers use specially treated aluminum cylinders (Wechter, 1976) that reduce these effects for certain gases but do not eliminate the problem, especially at low concentrations (Figs. 3 and 4).

Preconditioning, or pickling, the cylinders is another approach employed by many specialty gas suppliers to reduce the instability of gas mixtures. One common method of accomplishing this is to fill the cylinder with a gas mixture containing a higher concentration of the reactive component than will exist in the final mixture and to let it stand for several days. Some of the reactive material will react with, or be absorbed into, the wall surfaces, thus reducing the instability of the final gas mixture. This procedure helps reduce the problem but often does not totally eliminate it, especially at parts-per-million concentrations.

For these reasons, dynamic calibration standards may provide better accuracy when dealing with low concentrations of reactive or absorptive materials.

C. Dynamic Gas Mixing

The dynamic gas-mixing method is simply the dynamic blending or metering of two or more pure gases or gas mixtures. Several variations of the method are commonly used (Table II). For example, multiple stages can be used to dilute the desired component to very low concentrations (Lucero, 1976).

The dynamic method is most commonly used for preparing calibration standards at low pressures where they are needed for calibrating an analyzer. However, at least one specialty gas supplier uses this method for preparing some pressurized static gas standards. At pressures near atmospheric, the ideal gas law is applicable and good accuracy is attainable. [Refer to Eq. (8) in Section III,B,2,b to determine the composition of the gas mixtures.]

The dynamic method is especially useful for producing samples that contain components which tend to react with, or be absorbed

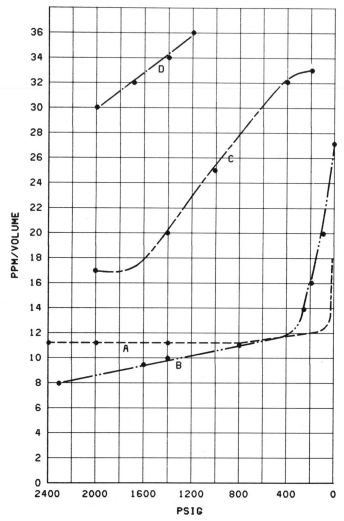

Fig. 3. Moisture standard concentration versus pressure. Curve A, aluminum cylinder; curves B–D, steel cylinders. [Adapted from Wechter and Kramer (1975) by permission of Instrument Society of America.]

by, the walls of a container that might be used to contain them. This can apply to nearly any compound when it is present at an extremely low concentration.

A large variety of equipment is now available to meter gases at total flow rates from less than 0.01 to more than 10 liters/min. This

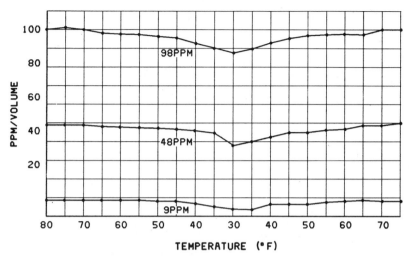

Fig. 4. Moisture concentration versus temperature of gas standard in aluminum cylinder. [Adapted from Wechter and Kramer (1975) by permission of Instrument Society of America.]

TABLE II

METHODS OF GENERATING GAS MIXTURES DYNAMICALLY

Method	Application	Advantages	Disadvantages
Pumps	Light, nonreactive gases	Good accuracy attainable	Expensive hardware; gases can react with or be absorbed by pump oil
Syringes	Adding trace components to a calibration mixture	Easy to determine composition of mixture; simple hardware	Diffusion of air into pure gas; syringe would require frequent refilling
Flow controllers	Primarily suitable for preparation of two- or three-component mixtures	Simple hardware; electronic types permit easy alteration of composition of mixture; flexible	Accurate flow measurement is sometimes difficult; must be calibrated for each gas
Permeation	Adding trace components; adding reactive components	Long life (sometimes several years); generate concentrations from ppb to ppm	Requires precise temperature control and good flow control of diluent gas; determining permeation rate is time consuming

(Table continues)

TABLE II (*Continued*)

Method	Application	Advantages	Disadvantages
Diffusion	Adding trace components; adding reactive components	Directly generate ppm concentrations; device is easy to fill	Requires precise temperature control and good flow control of diluent; determining diffusion rate is time consuming for lower concentrations
Exponential dilution	Generating a mixture with a continuously variable composition over a several decade range	Produces a predictable, continuously variable concentration of component of interest to very low (ppb) concentrations	Absorption of reactive components on internal surfaces causes errors; requires skilled technician
Gas saturation	Adding a single low–vapor-pressure component to a light gas	Can add compounds with low volatility at ppm to percentage concentrations; composition of mixture easy to determine from physical property data; diluent flow rate does not significantly affect accuracy	Requires precise temperature control; limited to one compound of low volatility in a gas mixture
Dynamic vaporization	Generating gas mixtures with several compounds of low volatility	Provides a gas mixture with several compounds of low volatility; mixture can be stored at low temperature; reactions are minimized	Capillary tubing can plug easily; vapor flow rate is difficult to measure

equipment includes piston pumps, pneumatic flow controllers, electronic flow controllers, rotameters, and pressure regulators with restrictors. A microprocessor-controlled gas-mixing instrument is available that permits the user to dial in the desired concentration of one or more components in a matrix. This instrument is precalibrated by adjusting the flow rate of each of the gases through a common orifice. The composition of the gas mixture is adjusted to the desired value and controlled by automatically operating solenoid valves to permit each gas to flow at the required average flow rate. The valves are cycled frequently, and a mixing chamber is utilized downstream of where the gases are blended to assure complete mixing. Each component can be selected at concentrations from less than 1 to 50%.

1. GAS-MIXING PUMPS

Piston pumps are available that can accurately meter two or more gases at pressures up to approximately 7 psig. These pumps are driven by a synchronous induction motor using intermediate interchangeable gears. Both the pump system and the gears are immersed in an oil bath. Mixing accuracies of 0.1% or better are claimed by one pump manufacturer.

The major limitation of the pump method is that the gases come into contact with the pump oil. Therefore, it is limited to use with light gases, such as hydrogen, air, or carbon dioxide, that do not react with the pump materials and are not dissolved in the oil.

2. MOTOR-DRIVEN SYRINGES

Motor-driven syringes are available that can meter one or more gases into a flowing gas stream (Nelson and Taylor, 1971). This method may be suitable for short-term experiments where maximum precision is not required. It is not applicable for continuous use over time periods of longer than 1–2 hr because of the limited capacity of suitable syringes and because of the diffusion of air into the syringe volume that would occur at low delivery rates. This method could be suitable for periodic introduction of a calibration standard over very long time periods if the syringe were automatically refilled with fresh material just before the desired calibration standard were needed. Under such conditions, the accuracy of the method could be within 2%. [Refer to Eq. (8) in Section III,B,2,b to calculate the composition of the gas mixture.]

3. MIXING WITH FLOW CONTROLLERS

a. Pneumatic flow controllers. Pneumatic flow controllers are available which can accurately meter gas flows from less than 1 ml/

min to over 100 liters/min. These devices operate by controlling a small differential pressure (e.g., 3 psi) across a restrictor through which the gas flows. Similar results can be obtained by using pressure regulators with restrictors. These devices can maintain their setpoints to within 1% for long periods of time (several months or years) if their temperature is controlled to within 1 or 2°C. [The composition of the blended mixture can be calculated by Eq. (8) in Section III,B,2,b.]

The major limitation of this method is that the actual flow rates must be determined by some other device, such as rotameters or soap-film bubble meters, which are limited in accuracy or are difficult to use. This method has been used, however, by such prestigious agencies as the NBS (Pella *et al.*, 1975).

b. Electronic flow controllers. Electronic flow controllers (Fig. 5) are available that not only control the flow rate but also indicate the flow rate on a digital display. These are feedback-type controllers which utilize a control valve to throttle the flow rate to a preselected setpoint. The flow rate is continuously variable over a range of almost two decades for each flow-control module. Modules are available that permit the control of flows from less than 0.5 ml/min to over 100 liters/min.

Each of these flow-control modules is accurate to within 1% of its full-scale range, even with ambient temperature changes from 5 to 40°C and upstream or downstream gas-pressure variations of 10 psi or more These features make these devices very convenient for use with many gas-blending applications.

The flow controllers are not true mass flow controllers so they must be calibrated for each gas. Response factors are available for most common gases, but the use of these factors may limit the accuracy to 4% of full scale. Also, it has been observed by the author that the 1% accuracy claimed by the manufacturers may not be met with factory calibration. Recalibration by the user may be required every few months to maintain maximum accuracy.

Operation of the flow controller is based on principles of heat transfer. When heat is applied to a gas stream, the temperature rise is a function of the mass flow rate, thermal properties of the gas, and the amount of heat applied. A small amount of heat added to the tube increases the temperature of a downstream temperature sensor compared with an upstream reference-temperature sensor as a gas flows through the tube. This temperature difference is directly proportional to the gas flow rate.

Fig. 5. Electronic flow controller.

4. Permeation Devices

A very useful and common application of dynamic gas mixing is the utilization of diffusion (Altshuller and Cohen, 1960) or permeation (O'Keefe and Ortman, 1966) devices. These devices (Fig. 6) permit the direct preparation of very low concentrations (parts per

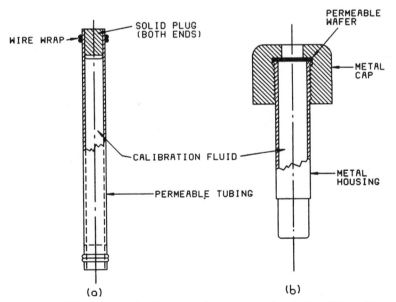

Fig. 6. (a) Permeation tube. (b) Low-emission permeation device. [Adapted from Baker and Brubaker (1974) by permission of Instrument Society of America.]

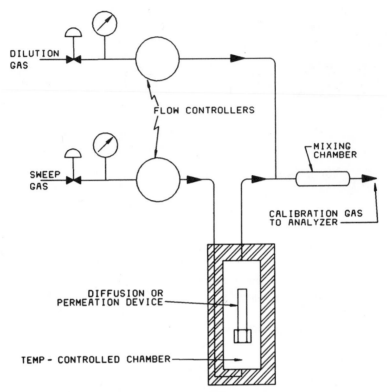

Fig. 7. Flow diagram for permeation or diffusion system. [Adapted from Baker and Brubaker (1974) by permission of Instrument Society of America.]

million or even parts per billion) of one or more components in a suitable matrix. One of these devices will generate a constant quantity of a desired compound for periods of time up to several years in many cases. Once the device is calibrated, a gas standard of known composition can be produced by flowing a diluent gas past it at a known and constant flow rate (Fig. 7).

A common permeation device is a Teflon tube (Fig. 6a), sealed at both ends, containing a compound that permeates the tube wall. Depending on the size and type of tube, permeation rates vary from less than 50 to more than 1000 ng/min at 30°C. These tubes are useful for providing calibration gases at concentrations of about 100 ppb to 10 ppm. Figure 6b illustrates a low-emission permeation device which can generate even lower concentrations.

The permeation rate of a gas through any permeation membrane is dependent on the type and size of the membrane, the differential partial pressure of the compound, and the operating temperature. Temperature is the most critical parameter. The permeation rate changes 10–15% for every Kelvin of temperature change for the commonly used Teflon membranes. A special type of permeation tube, described by Chand (1973), has a much smaller temperature effect of about 3%/K.

Permeation tubes are generally calibrated gravimetrically by determining the weight loss over a known time period (usually several weeks). Another method of calibrating the tubes, which requires much less time, is described by Dietz and Smith (1976); the method involves the determination of small pressure increases in a chamber of known volume.

The concentration of the vapor mixed with a diluent gas for either a diffusion or a permeation device can be determined by the equation

$$C = Rk/F, \tag{9}$$

where C is the vapor concentration in parts per million, R the permeation or diffusion rate in nanograms per minute, k the reciprocal density of the vapor in liters per gram, and F the total gas flow rate in milliliters per minute.

The value of k at the actual temperature and pressure at which the flow rate F is determined is calculated by the equation

$$k = \frac{22.4}{MW} \frac{T}{273.2} \frac{760}{P}, \tag{10}$$

where MW is the molecular weight of the vapor, T the absolute temperature in Kelvins, and P the pressure in torr.

5. DIFFUSION DEVICES

A diffusion device is ordinarily a liquid-filled reservoir with a long neck of precision-bore glass tubing (Fig. 8); however, any suitable materials that do not react with the liquid can be used. The liquid compound of interest is placed in the reservoir, the vapor from the liquid diffuses through the small-bore tubing, and a suitable diluent then mixes with the vapor as it is swept to the analyzer to be calibrated.

As with the permeation device, the quantity of vapor which diffuses from the diffusion device is constant if the device is held at a constant temperature and pressure. The tube is normally calibrated by determining the weight loss of the liquid over a known period of

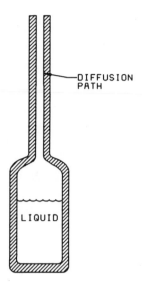

DIFFUSION PATH

LIQUID

Fig. 8. Diffusion tube.

time. Diffusion theory (Allshuller and Cohen, 1960) can predict the approximate diffusion rate, but the weight-loss method is used for maximum accuracy.

The larger the size of the reservoir, the longer the device can be used before replenishing the liquid. With a large enough reservoir the liquid could last for several years, and after the reservoir were depleted it could be easily refilled. This makes the method desirable for process analyzers that require frequent calibration checks. Figure 9 illustrates a long-life, high-rate diffusion device.

The rate of diffusion from a diffusion device is affected significantly by its temperature and pressure and the diameter and length of the diffusion path. Thus these factors affect the accuracy of the device. A temperature change of 1°C causes a change of approximately 5% in the diffusion rate, so it is very important to control temperature closely for good accuracy. Pressure changes have much less effect; an increase in pressure of 10% will cause the diffusion rate to be reduced by approximately 10%. The user should beware, however, of small pressure disturbances for very large diffusion devices. The actuation of a sample inject valve, for example, would tend to "pump" some of the concentrated vapor out of the reservoir, causing a temporary increase in concentration of the vapor in the diluent. In such cases it may be necessary to use a length of

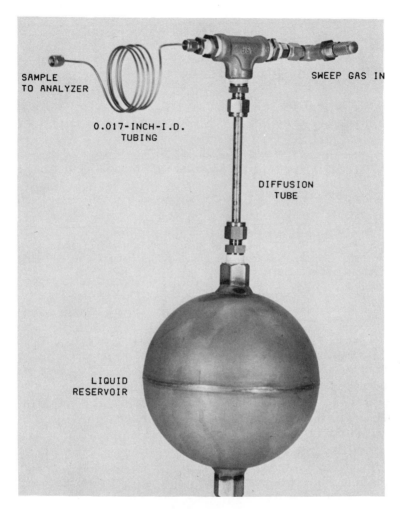

Fig. 9. High-rate diffusion device.

capillary tubing between the diffusion device and any possible disturbance.

Another factor to consider is the relative diameters of the liquid reservoir and the small-bore tube, especially if the tube is relatively short. If the liquid reservoir is relatively long and has a small diameter, then as the liquid is depleted the liquid level would affect the diffusion rate. The reservoir diameter should be at least 10 times the diffusion path diameter to minimize the influence of liquid level.

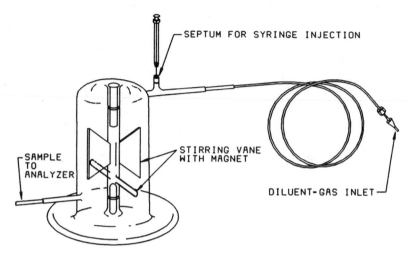

Fig. 10. Exponential-dilution flask. [Adapted from Hammarstrand (1976) by permission of Varian Instrument Group.]

6. EXPONENTIAL DILUTION

An exponential-dilution flask (EDF) is a simple device for calibrating detectors over a wide dynamic concentration range to the low parts-per-billion levels for many gases (Hammarstrand, 1976). A typical EDF (Fig. 10) consists of a 150–300-ml glass vessel with a diluent-gas inlet and an outlet that is connected to the instrument to be calibrated. A gas-tight syringe is used to introduce a known quantity of the pure sample gas to be employed for the calibration through the septum on the inlet side. The volume of the EDF and the diluent gas flow rate must be precisely known. By mixing the gases in the EDF with the stirring vane, an exponential dilution of the sample gas in the diluent gas is achieved over time. The concentration at any elapsed time can be determined by the equation

$$C = C_0 \exp[-(Q/V)^t], \tag{11}$$

where C is the concentration of the component of interest from the EDF in parts per million by volume, C_0 the initial concentration of the component in the EDF, Q the diluent flow rate in milliliters per minute, V the internal volume of the flask in cubic centimeters, and t the time elapsed after $C = C_0$.

In practice, the EDF can easily be set up to give a decade dilution of the initial sample gas concentration approximately every 10 min.

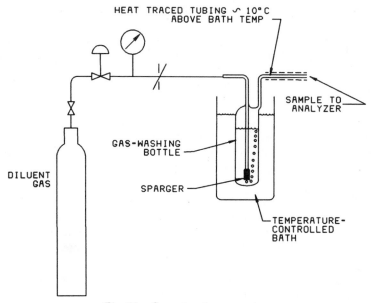

HEAT TRACED TUBING ∿ 10°C
ABOVE BATH TEMP

SAMPLE TO
ANALYZER

DILUENT
GAS

GAS-WASHING
BOTTLE

SPARGER

TEMPERATURE-
CONTROLLED
BATH

Fig. 11.   Gas-saturation apparatus.

At any desired time, the effluent from the EDF is sampled by the instrument being calibrated, and the results are compared with the calculated concentrations.

Ordinarily, some of the component of interest is adsorbed on the internal surfaces of the EDF at the beginning of a calibration run. As the concentration decreases, the adsorbed material is desorbed back into the diluent-gas stream. At very low concentrations, this quantity of desorbed material becomes a significant factor, causing errors in results.

The author has found that the EDF may be useful in characterizing the responses of detectors or analyzers over wide concentration ranges, but significant errors can result because of the adsorption problem with some compounds. Also, a high degree of skill is required to utilize the EDF in a manner that will produce accurate results even over small concentration ranges.

7. GAS SATURATION

Gas saturation is a simple method for producing known concentrations of vapors in a continuously flowing gas such as nitrogen (Fig. 11).

One version of Raoult's law (Perry, 1950) states that for ideal solutions the concentration of any component in the vapor can be determined from the vapor pressure of the pure component, the mole fraction of the component in the liquid, and total pressure by the equation

$$Y_A = P_A X_A / P, \tag{12}$$

where Y_A is the mole fraction of component A in the vapor, P_A the vapor pressure of pure component A at the same temperature, P the total pressure, and X_A the mole fraction of component A in the liquid.

This relationship can be simplified and results made more accurate by choosing a diluent gas that is relatively insoluble in the component of interest. For example, nitrogen is only about 0.002% soluble in isopropanol at 40°C. In that case, X_A becomes essentially equal to one.

The accuracy of this method is affected by the purity of the liquid, the reliability of the vapor-pressure data (Fig. 12), the stability of the saturator temperature control, the efficiency of diluent-gas saturation, gas compressibility, saturator pressure, and the care taken to prevent any condensation of the vapor produced. Temperature is the most critical parameter determining the accuracy and stability of the method. A temperature change of 1°C will change the concentration of the vapor by 5 to 10%. The author has found the overall accuracy of the method to be within 2% for several applications.

A significant advantage of this method over most other dynamic methods is that the accuracy and concentration of the mixture are not affected by the diluent flow rate at relatively low flow rates. This advantage is lost if the outlet gas is diluted by a second diluent gas at a point downstream of the saturator, as is recommended by some users (Paulsell, 1976).

This method is generally suitable for only one vapor at a time. If two or more liquids are used in the saturator, liquid composition will change with time, causing a change in vapor concentration.

A variation of this method is the utilization of saturated aqueous solutions of various crystalline substances to generate gas samples of constant humidity (Weast, 1969). One of these solutions at a constant temperature will produce a gas mixture of known and constant water-vapor percentage.

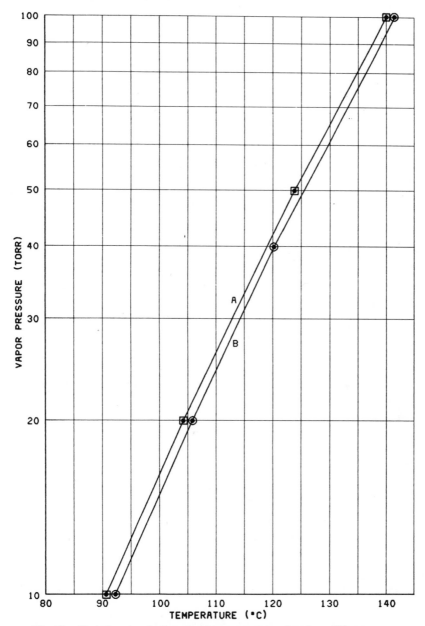

Fig. 12.   Variations in ethylene glycol vapor-pressure data from different sources. Curve A, Riddick and Bunger (1970) (by permission of Wiley-Interscience, New York); curve B, Stull (1947) (by permission of the American Chemical Society).

8. DYNAMIC SAMPLE VAPORIZATION

The dynamic sample-vaporization (DSV) method is useful for calibrating or checking analyzers measuring two or more components that would condense out of the sample at normal ambient temperatures and relatively high concentrations. The DSV method can also be used where only one condensable component is measured; however, another calibration method, the gas-saturation method (Section III,C,7), is normally more convenient for such an application.

The DSV method (Fig. 13) requires the preparation of a liquid mixture containing the components of interest in a suitable solvent. The mixture is then loaded into a container that can be pressurized to approximately 10 psig with a gas essentially insoluble in the liquid (such as nitrogen). The liquid is forced through a restrictor into a heated chamber, which serves as a vaporizer and mixing chamber. The mixed vapor exits from the vaporizer to the analyzer and is then routed to a collecting vessel or to a safe vent.

This method has the advantage that a stable mixture, one fairly representative of the actual process stream of interest, can be prepared. The method does not require the heating of a compressed gas cylinder to ensure that the components are mixed and contained in the vapor state. Such procedures are difficult to implement, and the danger of component reactions is greatly increased at higher temperatures. The DSV method permits the liquid sample to be kept cold, if necessary, until it is metered into the vaporizer. The vaporized sample is kept hot for only a few seconds before it is analyzed, reducing the chances of reaction and erroneous results.

There are some difficulties with the method, although the author has found it to be most convenient for many applications. The major difficulty is that residues, which normally accompany liquid mixtures, may plug the restrictor used to limit the liquid flow rate or plug the tubing entering the vaporizer. Another difficulty is that polymers can form, especially during vaporization, and this can eventually plug the vaporizer. It is difficult to determine when plugging occurs, unless a sensitive flow indicator (such as a rotameter) is used to monitor the vapor flow rate. The flow indicator itself could cause maintenance problems since it would have to be located in a heated zone.

To minimize these problems, proper care must be taken in designing and implementing this method. For example, a low-volume filter should be used upstream of the restrictor to minimize the buildup of residues in critical areas. The tubing used to connect the restrictor to

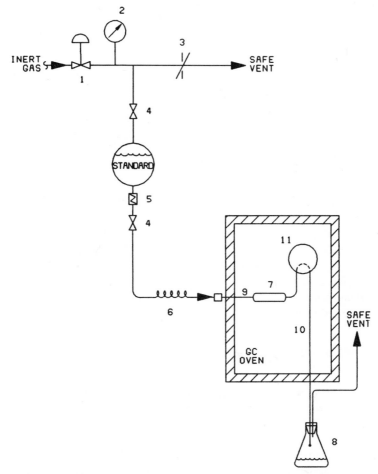

Fig. 13. Dynamic sample vaporization: 1, pressure regulator; 2, pressure gauge; 3, needle valve; 4, shutoff valve; 5, 25-μm filter; 6, capillary tubing; 7, vaporization chamber, 2-in.-long by 0.5-in. i.d.,; 8, collection vessel; 9, straight section of $\frac{1}{16}$-in. o.d. tubing (keep short for easy cleaning); 10, $\frac{1}{8}$-in. o.d. stainless-steel tubing; 11, sample inject valve.

the vaporizer should be of minimum length in the heated zone so that vaporization will occur in the vaporizer rather than in the tubing. However, this is difficult to achieve in practice. Some vaporization will likely occur in the tubing entering the vaporizer, and this is a predictable place for residues to form also. If properly designed, the tubing can be unplugged by disconnecting it at a fitting outside the heated zone and rodding it out with a length of wire or other suitable implement.

The liquid standard can be prepared on either a weight or a molar basis; the resultant vapor will be characterized on the same basis, since the liquid is totally vaporized. If the vaporizer is not sufficiently hot, however, some of the less-volatile components will not vaporize completely, resulting in a change in the vapor composition. The temperature of the vaporizer and any transfer tubing to the analyzer should be at least 10°C above the dew point to assure complete vaporization. [Refer to Cole and Reger (1969) for a method of estimating dew points.]

A typical liquid-sample flow rate to the vaporizer is 0.2–0.4 ml/min. For most organic liquids this would result in a vapor flow rate of approximately 50 to 150 ml/min. A sample flow rate that is too low will cause long response times. A sample flow rate that is too high will cause the vaporizer to flood and may also cause pressurization of the sample cell or sample loop used in the analyzer, thus causing calibration errors.

The author has found that accuracies within 2% can be achieved with this method.

9. Exotic Methods

Several rather exotic methods have been used to prepare dynamic gas mixtures. These methods, some still experimental, have been used primarily for highly reactive or unstable gas mixtures.

a. Photolysis of oxygen. One method that is routinely used to produce ozone calibration standards utilizes an ultraviolet lamp to generate ozone by the photolysis of oxygen (Hodgeson *et al.,* 1971). The quantity of ozone generated is controlled by the quantity of oxygen in the feed gas and also by the amount of radiation to which the oxygen is exposed. The amount of radiation is controlled by partially shielding the ultraviolet source from the gas stream with a movable sleeve mounted on a micrometer head. Ozone concentrations from a few parts per billion to several hundred parts per million can be generated by this method.

The quantity of ozone generated under specific conditions and parameters must be determined by some analytical method. One method is to titrate the ozone in the gas stream with a buffered potassium iodide solution (Byers and Saltzman, 1958).

b. Gas-phase titration.   Another exotic method that has been applied fairly routinely is the gas-phase titration of nitric oxide with ozone (Rehme *et al.,* 1972). By this method a sample containing nitric oxide is partially reacted with ozone, and the reduction in the concentration of nitric oxide is measured with a suitable analyzer (such as a chemiluminescent analyzer). Assuming that the reduction in the nitric oxide concentration results from the formation of nitrogen dioxide, the change in nitric oxide concentration is the same as the nitrogen dioxide produced. The ozone used for the titration is generated by an ultraviolet lamp as described in Section III,C,9,a. Calibration accuracy within 2% can be obtained with this method.

c. Electrochemical generation.   Reaction components such as ozone, nitric oxide, chlorine, and hydrogen sulfide have been electrolytically generated at the parts-per-billion to parts-per-million concentration levels in an air diluent stream (Harman, 1976). Theoretically, it is possible to accurately determine the quantity of each of these generated compounds by measuring the electrolysis current, but this cannot always be done in practice. Nevertheless, the method can be applied to the generation of certain reactive compounds if their actual concentrations are determined by some alternative method.

D. Static Liquid Mixtures

1. ORDINARY LIQUIDS

Ordinary liquid calibration mixtures can be prepared simply by adding the desired weight or volume of each liquid into a suitable container. As with gas mixtures, the accuracy of the mixture will be a function of the purity of the various liquids, the method of measuring the quantity of each liquid, and the stability of the mixture. Many liquid compounds are difficult to obtain with purity better than 95%, so significant errors can result, especially if the purity is unknown. Another source of error can exist if the analyzer is calibrated on the weight basis and the density of the sample to be analyzed varies significantly (see Section II,A,2).

2. LIQUID MIXTURES CONTAINING GASES

One type of liquid mixture is difficult to prepare: that which contains gaseous components at normal ambient temperatures. Such a mixture could be prepared by using the same gravimetric method described for the preparation of gas standards (Section III,B,2,b). However, unless the mixture is kept under sufficient pressure, some of the normally gaseous components will not be in the liquid phase. Also the composition of the liquid phase will change as liquid is depleted from the cylinder. It may be necessary to utilize a variable-volume liquid cylinder with a gas-driven piston to keep the mixture pressurized and to avoid the errors that would result from a two-phase mixture.

3. REACTIVE LIQUIDS

Some liquid streams contain combinations of chemicals that will react under certain conditions. Calibration standards that contain all the compounds might also react, especially if stored for several months. Reducing the temperature of the mixture will ordinarily reduce the rate of reaction, but this may not always work or be practical. It may be necessary to prepare fresh standards and use them immediately.

Another approach is to avoid adding the more-reactive compounds to a calibration standard containing a large number of compounds. The reactive compounds might be mixed individually in an inert solvent and remain stable. The use of two or more calibration standards might be an acceptable solution to these problems.

Compounds that are extremely reactive may require calibration by some method not requiring the introduction of a calibration standard (see Section II).

4. SUBSTITUTING A GAS PHASE

Occasionally, a problem arises where a liquid stream that contains one or more gases with high vapor pressures is being analyzed. It may be very difficult to prepare and utilize a liquid calibration standard containing these gases. If the analyzer is a GC, one solution to the problem is the utilization of a gas standard that contains the gas of interest, in either a pure state or a convenient matrix such as nitrogen. It may be necessary to use a second sample inject valve for the gas standard. The following equation was derived by the author for calculating the required concentration of each gas to be calibrated:

$$C_G = \frac{6.23 C_A V_L D T}{V_G (MW)_A P} , \qquad (13)$$

where C_G is the mole percentage of compound A in the gas sample; C_A the parts per million by weight of compound A in the liquid sample; $(MW)_A$ the molecular weight of compound A; D the liquid sample density in grams per milliliter; V_L the liquid-sample volume in microliters; V_G the gas-sample volume in microliters; P the absolute gas pressure in the sample valve (ordinarily barometric pressure) in torr; and T the gas-sample–valve temperature in Kelvins. The constant 6.23 includes factors for converting to standard temperature and pressure and for expressing results in mole percentage. For example, assume that the component to be calibrated in the liquid stream is ethane under the following conditions: $C_A = 1000$, $(MW)_A = 30.07$, $D = 0.6$, $V_G = 100$, $P = 762$, $T = 50°C$, and $V_L = 1$. Then, applying Eq. (13),

$$C_G = \frac{(6.23)(1000)(1)(0.6)(50 + 273.2)}{(100)(30.07)(762)} = 0.527\%.$$

It is difficult to determine the exact sample volumes. Therefore, the best accuracy would be obtained by using the same sample valve for both the liquid and gas samples. If this is done, V_G and V_L can be ignored since they cancel each other out.

If different sample inject valves are used, the ratio of V_L to V_G can be determined by comparing the response obtained when a standard is injected by both sample valves. To minimize errors, the detector should be linear over the operational range.

Another point of caution is that any change in liquid-sample density will cause error in the analysis because a fixed *volume* of sample is injected rather than a fixed *weight* (see Section II,A,2).

E. Dynamic Liquid Mixing

Most of the methods described for the dynamic mixing of gas standards could theoretically be used for liquid standards also, although only a few of the methods have been reported in the literature or used by the author. This is possibly because a crucial need does not exist. The dynamic-vaporization method or the gas-saturation method could not, of course, be applied to liquid samples. Also, the permeation and diffusion methods could not be used, although a similar approach, the use of a dialysis membrane (Tuwiner, 1962), might be used to transfer small quantities of one liquid component into another liquid stream. Only a few methods for dynamic liquid mixing will be discussed below.

1. Metering with Pumps

The use of pumps, such as peristaltic or piston, is probably the most common method for blending two or more liquid flows. Peristaltic pumps, in particular, are used for the precise metering of several liquids at low flow rates into a common mixture. These pumps can deliver flows from less than 1 to more than 100 ml/min. Although they are not commonly used to blend calibration mixtures, they can be used for that purpose. Peristaltic pumps have the capability of metering flows with an accuracy within 1% over a period varying from several hours to several weeks, depending on the flow rate and the nature of the liquid.

Some problems may be experienced when using peristaltic pumps. The elastomer tubing used in the pumps has a relatively short life, usually less than one month. At higher flow rates, the elastomer tubing wears out even faster. Also, some liquids attack the tubing, which shortens its life further and, in addition, could affect the composition of the blended mixtures. Another possible problem is that the flow rate cannot easily be adjusted to the precise values desired.

Piston pumps do not have problems associated with elastomer tubing, but they have other problems. For example, residues may damage the pumps. Also, the mixture composition may vary considerably over short time periods because of the pulsing action of the pistons.

2. Metering with Syringes

The method of metering with syringes can be applied to liquids in a manner similar to that used for gases (see Section III,B,2). In fact, the problem of air diffusion, which can occur with gases, is not nearly so serious for liquids. Accuracies within 1% are attainable with this method (Nelson and Taylor, 1971).

3. Metering with Flow Controllers or Pressure Regulators

Pneumatic flow controllers are available that can accurately deliver constant flow rates of clean liquids in ranges from less than 1 ml/min to several liters per minute. Two or more of these can be used to blend two or more liquids. The devices work by controlling a small differential pressure (e.g., 3 psi) across a restrictor through which the liquid is flowing.

The same type of metering can be obtained by utilizing pressure regulators and restrictors. However, flow controllers are normally better for low flow rates because they operate at lower differential pressures. This means that larger-diameter restrictors can be used, minimizing errors resulting from the buildup of residue in the restrictors.

The author is not aware of any electronic flow controllers for liquid streams, such as those available for gas streams. This may explain why the blending of liquid streams for calibration of analyzers is rarely done. With the other methods available, there may not be a critical need for dynamically metering liquid flows.

IV. Methods for Introducing Calibration Standards to Analyzer

A. Important Considerations

One can choose the best-possible calibration method and use the most-accurate calibration standard available and still obtain poor results if proper care is not taken in introducing the standard to the analyzer. For many process analyzers, the method of introducing the calibration standard is part of the sampling system. If good results are expected from the analyzer, then it is very important to utilize a reliable sampling system that will deliver samples representative of what exists in the process being monitored. An extensive discussion of sampling systems is beyond the scope of this chapter, but comprehensive treatment of this subject is given by Houser (1972) and Cornish et al. (1981).

Some methods of feeding calibration standards to analyzers that have been found useful by the author are described in this section. For simplicity, it is assumed that the standard is introduced to the analyzer by means of a sample inject valve. This valve contains a port or an external length of tubing that transfers a fixed volume of sample into the analyzer each time the valve is actuated. (Each time the valve is deactuated, the sample port or loop will be refilled with fresh sample.) Most methods discussed, however, will also apply to analyzers such as spectrometers that do not use sample inject valves.

B. Gas Mixtures

Normally, there are only two basic methods for introducing any sample into an analyzer: either push or pull it through. For gas samples, the most common method is to push it through. Even analyzers that utilize diffusion-type sensors usually require one of these basic methods for introducing calibration standards.

1. PRESSURIZED MIXTURES

Pressurized mixtures are easy to introduce into an analyzer. The simplest method is to connect a container of such a mixture to the sample inject valve with a length of tubing and to allow some sample to flow through the valve to a vent. As usual, the simplest method has some limitations. The quantity of gas standard available is usually limited, so some means of limiting the flow rate is needed. Also, if the sample flow rate is too high, the resistance to flow by the sample inject valve and vent tubing will pressurize the sample in the valve, causing an increase in the quantity of sample injected. Depending on the calibration method used, inaccurate data might result. Finally, many types of samples cannot be vented directly to the atmosphere because they would contribute to air pollution or be hazardous to the health of workers in the vicinity.

For these reasons, some means of controlling the sample flow rate should be provided. The flow rate should be fast enough to flush the sample inject valve thoroughly before each sample injection but not so fast as to cause a substantial buildup of pressure in the sample inject valve.

When utilizing dynamically generated mixtures, the flow rate can in some cases be controlled to a reasonable value without additional hardware. If the flow rate is too great, it may be necessary to vent the excess sample (Fig. 14).

When using pressurized static gas mixtures, it is sometimes sufficient to provide an adjustable restrictor between the sample source and the sample inject valve, although the addition of a pressure regulator at the source is usually necessary. A pressure regulator should always utilize some type of downstream restrictor to permit flow control at a reasonable pressure (Fig. 15). Too often a pressure regulator is used without a restrictor or with the restrictor adjusted to a wide-open position, so that the pressure regulator is not operating within its design parameters. Most pressure regulators will not regulate accurately at downstream pressures below 5 or 10 psig,

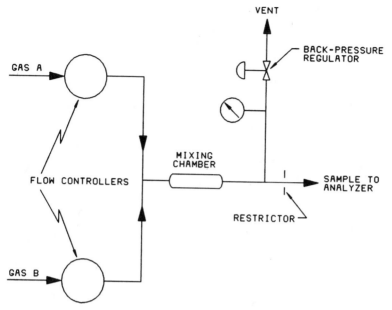

Fig. 14. Venting excess calibration gas.

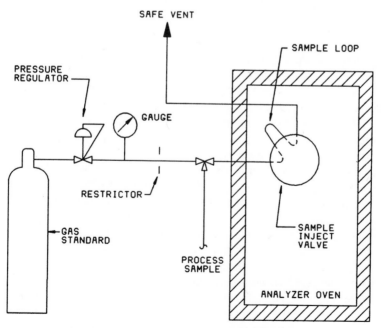

Fig. 15. Calibration with pressurized gas standard.

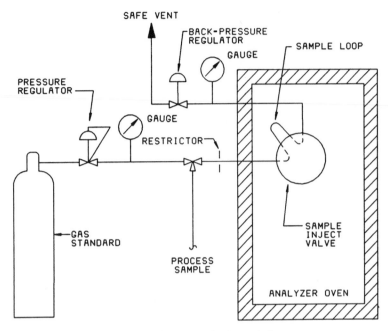

Fig. 16. Using pressurized-sample loop.

especially if the upstream pressure varies substantially. This will occur as the gas in a pressurized cylinder becomes depleted. Even if the regulator is capable of regulating pressure adequately at lower pressures, the sample flow rate could still vary drastically with even small changes in downstream restriction. This might occur when the sample inject valve is actuated, for example.

Any pressure regulator used should be constructed of materials that do not react with or absorb any of the components in the sample. Metal-diaphragm–type regulators should ordinarily be used. Polymer diaphragms absorb many compounds and also allow air to permeate into the sample stream.

When the sample must be vented from the sample inject valve into a header or a system that is not at a precisely controlled pressure (such as atmospheric), the problem becomes more difficult. (Of course, this problem would also have to be solved for the analyzer to provide reproducible data from a process stream). One solution is to operate the sample inject valve at a controlled pressure that is greater than the system pressure (Fig. 16).

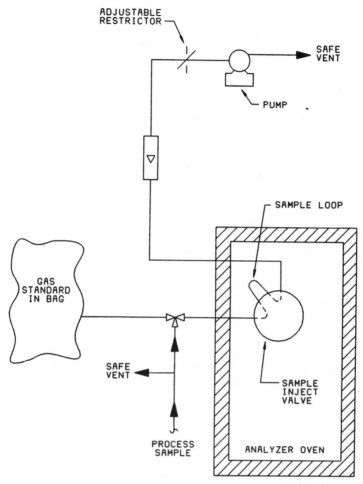

Fig. 17. Calibration with nonpressurized gas standard.

2. NONPRESSURIZED MIXTURES

Static mixtures at atmospheric pressure, such as those contained in a plastic bag, will require a pump either upstream or downstream of the sample inject valve. Sometimes a pump is already installed upstream of the valve because it is used as part of the analyzer's sampling system. Unless a pump is already installed, it is usually preferable to locate it downstream of the sample valve (Fig. 17). In the downstream position, a pump cannot alter the sample or contribute to the system lag time. A restrictor should be installed between the pump and the valve to prevent any pressure pulses caused by the pump from affecting the pressure in the sample inject valve.

Typical calibration flow rates are in the 50–100-ml/min flow range. (Analyzers other than GCs may require much higher flow rates.)

C. Liquid Mixtures

Liquid mixtures are often more difficult to handle than gas mixtures because they tend to contain more residues, which can plug sampling-system components or damage sample inject valves. Analyzers that handle liquid mixtures often require more maintenance as a result of this problem.

The simplest method of introducing a liquid standard is to pour it into a container mounted above the sample inject valve. The mixture is then allowed to flow by gravity through connecting tubing, through the valve, and then to a drain. Of course, this method has its limitations. First, if the mixture is exposed to the atmosphere, some of the components may evaporate, changing the sample composition and releasing possibly harmful vapors into the atmosphere. Second, the entire standard might be depleted in only a few minutes unless some method is used to restrict the flow rate. Finally, if the flow rate is too low, some of the components may vaporize while the standard is flowing through the sample inject valve, especially if the valve is heated or is located in a heated compartment.

Figure 18 illustrates a method that can be used to avoid these problems for many different types of liquid mixtures. The standard can be contained in a cylinder with fittings on the top and bottom. The top of the cylinder is pressurized with a gas such as nitrogen; the bottom contains a low-volume filter to remove sediment from the sample before it flows to the sample inject valve. An automatic shutoff valve, programmed to open a few seconds before the sample inject valve is actuated, is installed downstream of the sample inject valve. This will allow approximately 0.5–1 ml of fresh liquid to flow rapidly through the sample inject valve, fill the sample port, and flush out any gas bubbles that might have accumulated since the last sample injection. This method permits conditions to be optimized so that premature vaporization of the mixture is avoided while consumption of the standard is minimized. Also, it permits sealing off the standard so it will not evaporate.

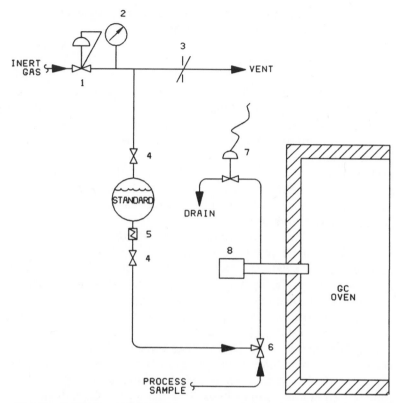

Fig. 18. Calibration with liquid mixtures: 1, pressure regulator; 2, pressure gauge; 3, adjustable restrictor; 4, shutoff valve; 5, 25-μm filter; 6, selector valve; 7, automatic valve that opens a few seconds before each sample injection; 8, sample inject valve.

Some liquid standards that contain highly volatile components may require a commercially available variable-volume cylinder with a gas-driven piston to prevent the alteration of the liquid sample composition as the mixture in the cylinder is depleted.

D. Vaporized Liquid Mixtures

Liquid mixtures that are to be dynamically vaporized before being injected into the analyzer can be fed to the analyzer or sample inject valve as shown in Fig. 13. See Section III,B,8.

V. Guidelines for Selecting Calibration Methods
and Standards for Field Applications

A. *Important Considerations*

The accuracy required of an analytical measurement system should have considerable influence on the choice of calibration method. Some analyzers, for example, are not required to provide accurate, quantitative data but are used to detect "breakthroughs" of one or more compounds into a process stream. Other analyzers may be expected to provide trend data, to reduce the severity of process upsets, etc., but are not required to be highly accurate. These analyzers would not require elaborate calibration methods for frequent field calibration.

Other analyzers may be required to provide data that will be used to control chemical processes at high levels of efficiency and also to produce accurate data for material balances, etc. Still other analyzers may be used to reliably detect the presence of toxic compounds for personnel protection. These analyzers might require the installation of hardware that will permit frequent automatic calibration of the analyzers in the field.

For these reasons, it is important to consider the field-calibration requirements in addition to initial-calibration methods before an analyzer is purchased or even specified. Analyzer measurement parameters sometimes can be chosen so that a calibration device may be mounted within a temperature-controlled oven already required by the analyzer, saving the cost of an additional oven. Calibration requirements should be discussed with the vendor or analyzer manufacturer at the time measurement requirements are discussed.

The ES method is probably the first that should be considered for the calibration of any process analytical instrument. It can be applied to virtually any analyzer by using one of the many different types of calibration standards available.

A second method of calibration, or at least a second type of calibration standard, should be used when first checking out the analyzer to verify the validity of the primary method. This will minimize the possibility of gross calibration errors. For applications requiring maximum accuracy, the use of a second method will provide an indication of the degree of accuracy attainable. The normalization method is a good second method that can be applied to most modern GCs.

B. Inert, Noncondensable Gases

The ES method should be the first method considered for the calibration of inert gases. Pressurized static gas mixtures are ordinarily the easiest type of gas standards to use, especially if more than one component must be calibrated. One of the dynamic methods of standards preparation might be considered if only one or two components are being measured. For example, electronic flow controllers could be used to meter two pure gases together.

C. Reactive Gases

For reactive gases, again, an ES method is recommended. Static gas mixtures may also be suitable for standards containing reactive gases, such as sulfur dioxide, especially if the concentration of the reactive gas is not extremely low. At lower concentrations, a dynamic mixture may be a better choice. A permeation device, for example, would be suitable for generating low concentrations of sulfur dioxide.

D. Condensable Gases

The ES method is a good choice for condensable gases. Static gas mixtures may be suitable as calibration standards if the condensable components are present at fairly low concentrations. A dynamic gas mixture, generated by the use of permeation or diffusion devices, may be preferable for extremely low concentrations. The gas-saturation method should be considered if only one condensable component is to be calibrated at high concentrations. If several condensable components at high concentrations must be calibrated, then the DSV method should be considered. More than one type of calibration standard may be necessary when the analyzer is measuring a combination of condensable or reactive gases and noncondensable gases.

E. Liquids

The ES method is a good choice for analyzers measuring liquid components. The normalization method should be used in conjunction with the ES method, especially if the results are desired on a weight basis. The ES would then be used to periodically check and/or adjust the response factors for each measured component.

Static liquid mixtures are normally the best type of ES. The utilization of a gas mixture might be useful for the calibration of highly volatile components (see Section III,D,4). A static liquid mixture could be used for calibration of the other components.

VI. Factors that Can Contribute to Calibration Errors

A. General

Many factors that can contribute to calibration errors have been discussed in the preceding sections of this chapter. Some of these are reiterated here. Some additional factors are discussed, particularly pertaining to analyzer hardware malfunctions. Some of these factors are not directly responsible for calibration errors, but the end result is erroneous data that may lead to frequent recalibration. Some of the factors are important only if the ultimate in calibration accuracy is required; others could cause serious errors.

For example, process personnel will sometimes obtain laboratory analyses of their process streams and compare the results with data from their process analyzers. The results frequently do not agree, and the process personnel often tend to believe the laboratory results. The results from both the laboratory and the process analyzers may be correct in such cases. The discrepancy may result because the laboratory and the process analyzer have samples taken from different locations, with possibly different compositions. This can occur when the process sample is not thoroughly mixed. The sample could be composed of two phases, which creates an extremely difficult sampling problem. Further discussion of sampling problems can be found in the work of Houser (1972) and Cornish *et al.* (1981).

Another factor that can lead to poor analytical results is the use of impure gases, such as the carrier gas for a GC. If an impurity in the carrier gas is the same compound as one of those being measured by the GC, substantial errors could result.

B. Sampling System

1. LEAKS OR RESTRICTIONS

Leaks somewhere within an analytical system are a frequent cause of poor data. Partial or complete blockage of a sample line can be just as serious, and this problem occurs frequently with liquid sampling systems or when sampling dirty or corrosive gas streams. Leaks upstream of or within the sample inject valve may cause dilution of the sample with air or carrier gas. Restrictions downstream of the sample inject valve may produce serious calibration errors or poor repeatability. This could occur, for example, if moisture condenses and freezes in the sample vent.

Leaks in stream-selector valves in multistream sampling systems can result in cross-contamination of one or more streams by other streams. The calibration standard may be affected if it is one of these streams.

2. IMPROPER TEMPERATURE OF SAMPLING SYSTEM COMPONENTS

For some applications, the sampling system may require heating to prevent condensation (or freezing) of certain components. If a part of the system is too cold, condensation may occur. If it is too hot, some of the components may be involved in reactions such as polymerization. In either case, the sample composition would change. This kind of problem can be identified by comparing the response generated by a calibration standard introduced directly into the analyzer with the response generated by the same standard introduced through the sampling system.

3. MATERIALS OF CONSTRUCTION

Reaction of some components of the process sample or calibration standard may occur if the wrong choice of sampling-system material is made. Also, elastomer or plastic components, such as tubing, may alter the sample by allowing permeation of air into the sample, permeation of some of the sample components into the atmosphere, or absorption of one or more of the sample components. When this occurs, sudden temperature changes in the tubing can change the degree of absorption and hence the sample composition. These problems would not show up in normal leak-test procedures (see Section V,B,2).

C. Analyzer

1. DEFECTIVE HARDWARE

Poor analytical results will likely be obtained if any of the components of the analyzer, such as detector, temperature controller, columns, or valves, are defective. Most malfunctions of this type would affect the calibration standards in the same manner as the process sample. Such problems would be indicated by variable results generated by the calibration standards.

2. IMPROPER MEASUREMENT PARAMETERS

Sometimes poor analytical results are not caused by defective analyzer hardware but by nonoptimum measurement parameters. For example, the wrong type of GC column might not adequately separate two components in the sample. One of the components, which may not be of interest to the user of the instrument, may increase the indicated concentration of the measured component. Selection of appropriate gas standards can help identify such problems.

Analyzer manufacturers normally have service personnel who can correct analyzer problems. Instruction manuals provided by instrument manufacturers normally include information that can guide trained instrument technicians in locating and correcting analyzer problems for users who prefer to perform their own maintenance.

D. Other Significant Factors

1. BAROMETRIC PRESSURE

Barometric pressure is often ignored as a factor contributing to poor analytical results. However, barometric pressure directly affects the sample quantity in almost any analytical instrument that measures a gaseous sample. Most analyzers, for example, utilize a sample valve with a fixed-volume sample loop or sample cell. If the barometric pressure increases by 1%, the density of the gas increases by 1% (for an ideal gas), and therefore the response for each component will increase proportionally.

The normalization or IS methods of calibration would compensate for the effect of barometric pressure changes or any other factor that might affect the apparent sample size used by an instrument. Another method for eliminating this effect would be to use an absolute-pressure regulator to control the sample pressure at the analyzer. This added complication to the analyzer, however, may not be justified by the relatively small errors that are caused by barometic pressure changes. Only occasionally will the barometric pressure change by more than 1% over a 24-hr period. Changes in barometic pressure can also slightly affect the detector response for analyzers such as GCs. Therefore, even if the sample pressure is controlled, slight errors may result unless a calibration method such as the normalization method is utilized.

2. GAS COMPRESSIBILITY

The effect of gas compressibility on the accuracy of pressurized static-gas standards was discussed in Section III,B,2. However, even if the composition of the calibration standards is accurately and precisely known, some errors in analytical results can occur with complex gaseous mixtures containing one or more highly compressible gases (those which deviate from the ideal gas laws). These errors can occur if the concentrations of each of three or more components in a set of samples varies widely. In that event, the effective sample size may be different for each sample. For most applications, the error introduced would be very small (less than 2% of the contained amount of each component). The normalization method would compensate for any errors introduced by this phenomenon. However, this phenomenon could cause small errors in the determination of component response factors used with the normalization method.

If the normalization method cannot be used, any errors that might result from the variable compressibility of various gas mixtures can be minimized by utilizing calibration standards that are very similar to the composition of the actual stream being analyzed.

ACKNOWLEDGMENTS

The author greatly appreciates the assistance received during the preparation of this chapter from the following co-workers at Union Carbide: R. S. Berry, J. H.

Brubaker, J. F. Fisher, B. J. Meneghelli, J. R. Moss, D. R. Steele, and W. H. Wagner.

References

Altshuller, A. P., and Cohen, I. R. (1960). *Anal. Chem.* **32**, 802.

Baker, G. L., and Brubaker, J. H. (1974). In "Analysis Instrumentation" (W. V. Dailey, J. F. Combs, and T. L. Zinn, eds.), vol. 12, pp. 135–137. Instrument Society of America, Pittsburgh.

Byers, D. H., and Saltzman, B. E. (1958). *Am. Ind. Hyg. Assoc. J.* **19**, 251–257.

Chand, R. (1973). Improved permeation devices for calibration. Presented at 66th Annual Meeting of the Air Pollution Control Association, Chicago.

Cole, K. M., and Reger, J. A. (1969). In "Analysis Instrumentation," (B. L. Connelly, L. Fowler, and R. G. Krueger, eds.), vol. 7, p. 162. Instrument Society of America, Pittsburgh.

Cornish, D. C., Jepson, G., and Smurthwaite, M. J. (1981). "Sampling Systems for Process Analysers." Butterworth, London.

Crabtree, J. H., and Blum, D. (1981). *Instrum. Technol.* **28**(8), 41–42.

Dietz, W. A. (1967). *J. Gas Chromatogr.* **5**, 68–71.

Dietz, R. N., and Smith, J. D. (1976). In "Calibration in Air Monitoring," pp. 164–179. American Society for Testing and Materials, Philadelphia.

Draper, N. R., and Smith H. (1966). "Applied Regression Analysis." Wiley, New York.

Gordan, A. J., and Ford, R. A. (1972). "The Chemist's Companion," p. 487. Wiley, New York.

Guillemin, C. L. (1980). *J. High Resolut. Chromatogr. Chromatogr. Commun.* **3**, 620–623.

Hammarstrand, K. (1976). *Varian Instrum. Appl.* **10**(2), 14.

Harman J. N., III (1976). In "Calibration in Air Monitoring," pp. 282–300. American Society for Testing and Materials, Philadelphia.

Hodgeson, J. A., Stevens, R. K., and Martin, B. E. (1971). "A stable ozone source applicable as a secondary standard for calibration of atmospheric monitors." Presented at Analysis Instrumentation Symposium, Houston.

Houser, E. A. (1972). "Principles of Sample Handling and Sampling Systems Design for Process Analysis." Instrument Society of America, Pittsburgh.

Hughes, E. E. (1976). In "Calibration in Air Monitoring," pp. 223–231. American Society for Testing and Materials, Philadelphia.

Keulemans, A. I. M. (1959). In "Gas Chromatography," (C. G. Verver, Ed.), 2nd ed., p. 35. Van Nostrand-Reinhold, Princeton, New Jersey.

Kipiniak, W. (1981). *J. of Chromatogr. Sci.* **19**, 332.

Lee, W. G., and Paine, J. A. (1976). In "Calibration in Air Monitoring," pp. 210–219. American Society for Testing and Materials, Philadelphia.

Lucero, D. P. (1976). In "Calibration in Air Monitoring," pp. 301–319. American Society for Testing and Materials, Philadelphia.

Nelson, G. O., and Taylor, R. D. (1971). *Anal. Chem.* **43**(10), 1340–1342.

Nowicki, H. G., Devine, R. F., and Kieda, C. A. (1979). In "Measurement of Organic Pollutants in Water and Wastewater" (Van Hall, ed.), pp. 130–151. American Society for Testing and Materials, Philadelphia.

O'Keefe, A. E., and Ortman, G. C. (1966). *Anal. Chem.* **38**, 760.

Paulsell, C. D. (1976). *In* "Calibration in Air Monitoring," pp. 232–245. American Society for Testing and Materials, Philadelphia.

Pella, P. A., Hughes, E. E., and Taylor, J. K. (1975). *Am. Ind. Hyg. Assoc. J.* **36**(10), 755–759.

Perry, J. H. (ed.) (1950). "Chemical Engineers' Handbook," 3rd ed., p. 317. McGraw-Hill, New York.

Rehme, K. A., Martin, B. E., and Hodgeson, J. A. (1972). The application of gas-phase titration in the simultaneous calibration of NO, NO_2, NO_x, and O_3 atmospheric monitors. Presented at 164th ACS Meeting, New York.

Riddick, J. A., and Bunger, W. B. (1970). *In* "Organic Solvents," p. 149. Wiley, New York.

Roccanova, B. (1968). The present state of the art of the preparation of gaseous standards. Presented at the Pittsburgh Conference on Analytical Chemistry and Spectroscopy, Cleveland.

Roy, C., Bellemare, G. R., and Chornet, E. (1980). *J. Chromatogr.* **197**(2), 121–127.

Stedman, D. H., Kok, G., Delumyea, R., and Alvord, H. H. (1976). *In* "Calibration in Air Monitoring," pp. 337–344. American Society for Testing and Materials, Philadelphia.

Stull, D. R. (1947). *Ind. Eng. Chem.* **39**, 517.

Tuwiner, S. B. (1962). "Diffusion and Membrane Technology," p. 10. Van Nostrand-Reinhold, Princeton, New Jersey.

Weast, R. C. (1969). "Handbook of Chemistry and Physics," 50th ed., p. F–40. The Chemical Rubber Company, Cleveland, Ohio.

Weast, R. C., Selby, S. M., and Hodgman, C. D. (1964). "Handbook of Chemistry and Physics," 45th ed., p. F–42. The Chemical Rubber Company, Cleveland, Ohio.

Wechter, S. G. (1976). *In* "Calibration in Air Monitoring," pp. 40–54. American Society for Testing and Materials, Philadelphia.

Wechter, S. G., and Kramer, F., Jr. (1975). Evaluation of gas phase moisture standards prepared in treated aluminum cylinders. Presented at 21st Annual ISA Analysis Instrumentation Symposium, Philadelphia.

Williard, H. H., Merritt, L. L., Jr., and Dean, J. A. (1974). "Instrumental Methods of Analysis," 5th ed., pp. 379–380. Van Nostrand-Reinhold, Princeton, New Jersey.

14

Interfacing Analyzers to Computers

JOHN C. HUDELSON

Amoco Oil Company
Whiting Refinery
Whiting, Indiana

I. Introduction

In a large process plant it would be unwieldy to monitor processes by traditional methods; one would have to drive between jobsites and or make several phone calls to various operators to get an overall picture. Fortunately, one can connect analyzers to computer equipment and call up historical and current data on a terminal. However, several steps must be taken before this can come about.

The analyzer signal (electrical or pneumatic) must be connected to the computer without introducing excessive noise to the signal. The computer(s) must be programmed to read the signal and store data in bulk memory (typically magnetic disk). Finally, programs must be written to enable the computer to retrieve and display the data on a terminal (CRT or printer). If distances are great (such as in a refinery) and the CRT or printer is far from the computer, signals to and from the terminal must be changed to a form that can be transmitted over long distances and converted back to the original form at the other end.

Although extensive effort is required, it is worthwhile, since data-monitoring systems enable personnel to efficiently operate the plant with minimum energy consumption while ensuring that the product meets specifications.

II. Process Parameters and Instruments

This section is a brief review of the instruments used in industry. In a typical installation, signals from a unit are relayed to a local control room and connected to meters, trend recorders, alarms, etc., to present data to the operators. In this section we shall list measurements and instruments commonly used to monitor processes.

A. *Flow*

1. ROTOMETER

A common method of measuring flow is with a rotometer. In this device the gas or liquid enters a tapered tube from the bottom and flows to the top. A ball "rides" on the fluid as it flows to the top, and the more flow there is, the higher the ball rides in the tube. Calibrations are marked so that one may read the flow rate.

2. ORIFICE

A second method of measuring flow is with an orifice. A plate with a hole (orifice) is placed in the line. The pressure drop across the plate is measured, and the flow is calculated from the pressure drop. Unfortunately, energy is wasted in the pressure drop.

3. VENTURI

Venturi methods are used to decrease the loss of energy due to friction. The venturi (inserted in place of straight pipe) tapers down to a constriction and then gradually returns to the original diameter. The converging section converts the pressure head of the fluid to a velocity head, and the diverging section converts the velocity head back to a pressure one. Pressure readings are made between the original diameter (where pressure is at the maximum) and the constricted diameter (where pressure is at the minimum). The flow is calculated based on the pressure differential.

4. PITOT TUBE

Still another method makes use of the Pitot tube. A Pitot tube protrudes into the pipe, and the end of the tube is turned against the flow. The kinetic energy of the fluid is converted into pressure. The difference between the Pitot-tube pressure and the pipe pressure may then be measured by a pressure probe. Flow is calculated from the differential pressure.

Other methods of measuring flow include bellows (such as those used in utility gas meters), rotating vanes, and pistons (Soisson, 1975; Andrew and Williams, 1979).

B. Level

1. SIGHT GLASS

One method of measuring levels is with a sight glass. This may be a transparent tube that is connected to the bottom of a tank and runs up its side.

2. FLOAT AND HYDROSTATIC METHODS

Another method uses a float, which may be monitored by remote instrumentation. With the hydrostatic method, the level is measured by the pressure exerted on the bottom of the tank.

3. ACOUSTIC METHOD

Tank levels may be measured by acoustic methods. A pulse is sent out by a transmitter, reflected from the gas–liquid or air–liquid interface, and received. The tank level is determined from the elapsed time, which is proportional to the distance between the transducers and the tank level.

4. CAPACITANCE METHOD

Still another method uses capacitance. Since the liquid has a different dielectric constant from the gas above it, the capacitance between the tank wall and an interior electrode changes as the level in the tank changes (Andrew and Williams, 1979; Soisson, 1975).

C. *Pressure*

1. MANOMETER

The first instrument designed for measuring pressure was the manometer. A mercury barometer is a type of manometer for absolute-pressure application. A manometer has a U-shaped transparent tube partially filled with a dense liquid. The tube is connected across the pressure differential to be measured. The differential pressure supports a difference of liquid level across the tube. This instrument is not practical for high pressures or connection to remote instruments.

2. BOURDON TUBE

Typical pressure gauges used in industry are of the bourdon-tube type. The tube is coiled, but as pressure increases it begins to straighten. The tube is connected mechanically to a pointer, which indicates pressure on a meter face.

3. STRAIN GAUGES

Pressure may also be measured with strain gauges. A membrane is placed between two regions where the pressure difference is to be measured, and as pressure increases, the membrane stretches. A strain gauge (a form of resistor) is placed on the membrane. As the membrane stretches, the strain gauge becomes longer, and the cross section decreases, causing the resistance to increase. Pressure is determined from resistance readings (Soisson, 1975; Andrew and Williams, 1979).

D. Temperature

1. THERMOCOUPLES

When dissimilar metals are joined together, they develop an electrical potential that changes as a function of temperature. Regions where dissimilar metals are joined are called *junctions*. A practical thermocouple has a reference junction (kept at a known temperature) and a measuring junction (to measure the unknown temperature). Both junctions consist of the same two metals, Type 1 and Type 2. The two junctions are connected together by the Type 1 metal, say. Then a potential is measured across the two leads coming from the Type 2 metal at each junction. If a third metal, Type 3, is used to extend the Type 2 metal leads, the potential across the reference and measuring junctions remains the same if both Type 2–Type 3 connections are at the same temperature, since the voltages developed between Type 2 and Type 3 metals cancel out. The measured temperature is obtained from the net potential across each of the Type 2 metal leads and the reference temperature.

2. RESISTANCE TEMPERATURE DETECTORS

Changes of resistance as a function of temperature are used to measure temperature with resistance temperature detectors (RTDs) (Andrew and Williams, 1979; Soisson, 1975).

3. PYROMETERS

Pyrometers measure the infrared radiation from objects. Circuitry in the pyrometer converts the infrared radiation to a temperature reading by a modified version of the black-body radiation equation

$$W = eK(T^4 - T_0^4), \tag{1}$$

where W is the radiant energy from a unit area of a black body, e the emissivity constant, K the Stefan–Boltzmann constant, T the absolute temperature of the target, and T_0 the absolute temperature of the surroundings. Emissivity varies for different surfaces. It may range from zero to one. For this reason, pyrometers have a dial to adjust for different surfaces (Soisson, 1975).

E. Stream-Composition Process Analyzers

Since process analyzers are covered in detail in other chapters, this subsection is only a brief description of commonly used analyzers.

1. OXYGEN AND HYDROGEN SULFIDE ANALYZERS

Most analyzers, including oxygen and hydrogen sulfide analyzers, perform a continuous analysis of a process variable. Oxygen analyzers may be used to monitor oxygen in flue gas, so that the air–fuel ratio may be adjusted for efficiency and low pollution. They may also be used to detect the presence of oxygen in a hydrocarbon stream. Solid-state and paper-tape analyzers may be used to detect the presence of hydrogen sulfide down to a parts-per-million basis in air or in a hydrocarbon gas (Andrew and Williams, 1979).

2. PHOTOMETRIC ANALYZER

Photometric analyzers continuously measure the absorption of radiation at a selected wavelength. The wavelength must be chosen so that absorption will be maximized for the measured component and minimized for other components. Photometric analyzers may measure infrared to ultraviolet wavelengths (Soisson, 1975).

3. GAS CHROMATOGRAPH

An analyzer that does not perform a continuous analysis is the gas chromatograph (GC). In a process environment, a process GC cycles repeatedly. A sample is injected into a column, and the sample separates into components as it passes through the column, which is pressured by a carrier gas, and a detector senses the presence of the components. The peak areas or peak heights of the components are electronically measured. Then the composition of the stream is calculated from the peak areas or heights. There are trend-value outputs from most process GCs. These trend values are changed only after the end of each cycle. In addition to trend outputs (one trend per component, 4–20-mA current loop), some GCs have serial RS-232C ports so that they may be directly interfaced with a host computer (Andrew and Williams, 1979; Soisson, 1975).

III. Interfacing Analyzers to Computer Systems

A. Overview

Several steps are necessary before one can walk to a computer terminal and call up data concerning some remote process analyzers. One step is to establish the hardware interface of the process analyzer to the computer. Care must be taken to minimize noise interference. The following subsections describe different types of signals, signal conversions, and filtering to minimize noise. The next step is to write programs that will store the data and display it on a computer terminal when requested. A brief description of computer programming and computer operating systems is included later in this chapter.

B. Types of Signals

In process industries, so-called *live-zero signals* (where the zero of a measurement is a nonzero signal level) are used. If the signal level goes to zero, it is apparent that the equipment has failed (Liptak, 1970).

1. PRESSURE

Pressures of 3 to 15 psig are commonly used in the petroleum industry. A significant reason to use pneumatics is that they are intrinsically safe in explosive atmospheres (Liptak, 1970). In such atmospheres, all electrically operated devices must also be intrinsically safe (power limited to prevent ignition of the atmosphere), placed in a purged enclosure to prevent the buildup of hazardous vapors around the electrical equipment, or placed in a heavy enclosure designed to contain a possible explosion. Hence, in a potentially hazardous environment, pneumatic instruments may be less expensive than corresponding electrical instrumentation in some applications. A disadvantage of pneumatic instruments is that the lag time for a 63.2% complete response at a distance of 100 ft (30 m) is approximately 0.2 sec. Their response time increases with distance (Liptak, 1970).

2. ELECTRICAL-ANALOGUE SIGNALS

Many probes use low-level voltage signals. Thermocouples, thermistors, and RTDs are among the instruments that use low-level signals (100 mV or less). Of course, a significant disadvantage of such low-level signals is noise. If data are to be transmitted for long distances, the signal should be converted to another form.

If distances are not great and there is not much noise present, high-level (1–5-V) signals may be transmitted satisfactorily. The range of 1 to 5 V is commonly used for trend recorders and indicators.

For longer distances, current signals may be used. The most commonly used range is 4–20 mA. Several current-measuring devices may be connected in series to form a loop. Figure 1 shows a typical current loop. In a current loop, the resistance of the loop [including load resistors (R_{TR} and R_C) and wire and connector resistances] must not exceed the compliance limit of the current transmitter, or

$$R_{loop} \le V_{ct_{max}}/I_{fs}, \tag{2}$$

where R_{loop} is the total resistance of the loop, R_C the resistance across the computer input, R_{TR} the resistance across the trend recorder, $V_{ct_{max}}$ the maximum voltage the current transmitter can develop, and I_{fs} the full-scale current for the current loop. For example, in a 4–20-mA current loop, if $V_{ct_{max}}$ is 24 V, then the resistance of the current loop must not exceed 1200 Ω.

In process instrumentation, the analogue 4–20-mA signal continues to be in wide use despite the inroads of digital transmission methods. If a digital link fails, communication for several variables is lost. However, since there is only one variable per 4–20-mA link, transmission of other variables would continue on the remaining links if one were to fail (Andreiv, 1982).

3. COMMAND–STATUS BINARY SIGNALS

In addition to analogue signals mentioned above, there are command and status binary (two-state) signals. Such signals are either on or off, with no intermediate values. Traditionally, switches and relays are used to implement control of the signals. Applications include the control of motors, lights, and alarms.

4. SIGNAL CONVERSIONS

Signal conversions are performed to make the data signal more immune to noise or to present the data in a more convenient form. Signals may be converted from one type analogue to another, to sample-and-hold form, or by multiplexing–demultiplexing.

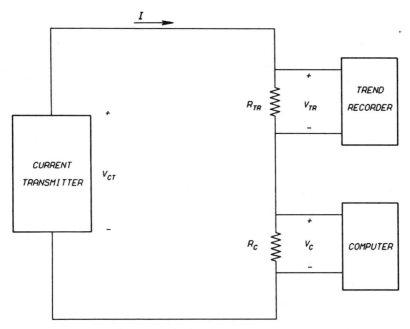

Fig. 1. Typical analogue current loop.

Low-level signals (millivolt range) are, of course, susceptible to noise. Such signals may be converted to a high-level voltage or, for longer distances, to a 4–20-mA signal. A resistor may be used at a voltage input to convert current to voltage. The voltage input in this case should be isolated; that is, neither side of the input should be connected to ground. Otherwise ground loops may result, where unpredictable current flow will take place between the transmitting and receiving instruments via ground. If greater noise immunity is desired, the voltage- or current-analogue signal may be converted to frequency. At the frequency output, the frequency may be converted to voltage or current, or the pulses may be directly counted in a certain time frame.

Another possible conversion is pressure to voltage or current. Most analogue inputs to computers are of the voltage or current type, but pneumatics are frequently used in instrumentation and control since they are intrinsically safe in explosive atmospheres.

Sample-and-hold conversions are used when signal values at certain times are to be retained and outputted as a constant until the incoming signal is sampled again and the output of the conversion device is changed to a new value. Such devices are useful for analyzers that run in cycles, where the result is to be displayed on trend recorders and indicators to operators in a control room (Liptak, 1970).

Several signals, including analogue and status–command, may be multiplexed (combined together), sent, and demultiplexed (original signal recreated) at the other end, as shown in Fig. 2. The advantage, of course, is that only two wires, instead of many, are needed to transmit many signals from one location. In Fig. 2 the original signals (designated in the figure as C/S_n for the command and status signals and A_n for the analogue signals, with contacts and current, respectively, as sources, where $n = 1$–3) are shown to the left of the multiplexer; to the right of the demultiplexer are the recreated signals corresponding to the originals.*

5. SIGNAL NOISE PROTECTION

Signals may be protected from noise by using shielded and grounded cable. However, to protect against ground loops the shielding should be connected at only one end.

Another method of protecting signals from noise is to use a twisted pair of conductors with a signal input with high common-mode rejection. The fact that it is a twisted pair ensures that both conductors pick up the same interference. A signal input with good common-mode rejection will reject the noise, since it is the same in both conductors, but will accept the desired signal, which appears as a difference between the two conductors (Liptak, 1970). A good input may have a 160-dB common-mode rejection rate (Analog Devices, 1981).

Low-pass filters may also be used to decrease noise. If a signal changes very slowly, a filter with a cutoff frequency of less than 1 Hz may be used, and such a filter would do an effective job of filtering out transients. Of course, such a filter could not be used for rapidly changing signals, such as peaks from a GC detector. For a rapidly changing signal, the filter must have a much higher cutoff frequency (Liptak, 1970). With a higher cutoff frequency, the signal channel is more susceptible to noise.

---

* Multiplexing and demultiplexing equipment with the tradename of Teleterm-Squeezer may be obtained from Conlog, Inc., P.O. Drawer 11001, Stafford, Texas 77477. In the minimum configuration, 16 status–command signals may be sent through the system. Expansion modules may be added to accommodate analogue and additional status–command signals.

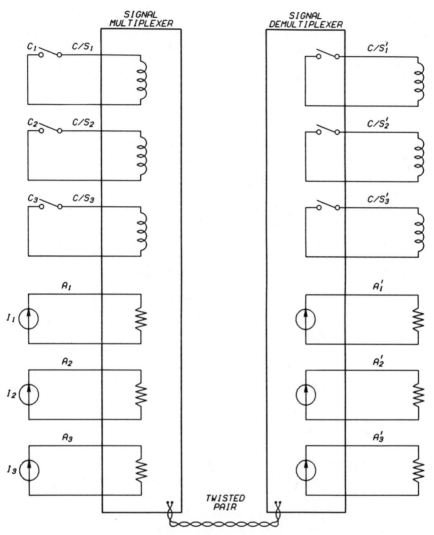

Fig. 2. Signal multiplexing.

6. SIGNAL ISOLATION

If a signal input is single ended (that is, one side of the signal is tied to ground) or if the common-mode voltage levels (steady state or transients) exceed the tolerance of the input, an isolation module may be placed between the signal and the input. There are two types of isolation modules. One is the passive module, which is powered by the primary current loop. The module will impose a comparatively high resistance (or voltage drop) on the loop. As discussed previously, the sum of the resistances of a current loop may not exceed the capabilities of the current source. For example, from Fig. 3

$$R_W + R_I + R_{MOD} \le R_{compliance} \qquad (3)$$

must hold, where $R_{compliance}$ is the maximum resistance that the primary current source can drive, R_W the wire resistance, R_I the resistance across an instrument input on the primary current loop, R_{MOD} the resistance across the isolation module input, R_{INS} the resistance across an instrument on the secondary current loop, i_P the primary loop current, and i_S the secondary loop current.

Since the device obviously requires power to operate, the maximum voltage drop across the secondary loop is less than the maximum voltage drop across the primary loop, hence

$$R_{INS} < R_{MOD} \qquad (4)$$

for the same current level on both sides of the module. To convert voltage to current, or if

$$R_{INS} \ge R_{MOD}, \qquad (5)$$

a powered isolator must be used (Conlog, 1980).

C. Signal Interfacing to Computer System

Analogue signals may be directly interfaced to a computer over moderate distances. For greater distances, where noise may become a problem, or where there are many signals to be picked up from a single location, analogue signals may be converted to a digital representation and transmitted over great distances. A single analog signal may be converted into a frequency representation, and the pulses may be counted by the computer in a specified time frame. To obtain good resolution, many pulses must be counted, which means that the data-acquisition program must remain active in the computer for the relatively long time needed to count the pulses.

If there are several signals at a remote location, a single-board microcomputer, with expansion boards as needed for extra inputs and outputs, may be installed at the location. The microcomputer will send coding representing the analog and command–status sig-

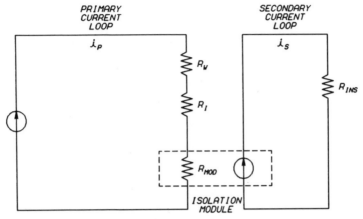

Fig. 3. Current-signal channel with isolation module.

nals upon request from the host computer. Error checking is done by parity and/or checksum methods (to be described later). Several dozen signals may be interfaced into a single-board micro and its expansion boards. Two dedicated phone circuits are needed to interface the microcomputer to the host computer. Figure 4 shows a schematic of a data-acquisition microcomputer connected to a host computer. The link between the micro and host computers may be RS 232C, RS 449, 20-mA digital current loop, line drivers, modems, or radio. The methods of communication will be described in the following sections.*

For applications where hard-wired links are cumbersome or totally impractical, such as communication with a moving system (e.g., a crane in a steel mill), signals may be transmitted by radio. Examples are the VR-20, VR-30, and VR-11 systems manufactured by Vectran Corporation.† The VR-20 system is suitable for transmitting 2 analog or 16 command–status signals between a transmitter–receiver pair. The VR-30 system may be used to send serial digital data between a transmitter–receiver pair. For more complex applications, the VR-11 system may be used, in which the master unit is connected to the host computer. The master unit may access many satelite units, each of which can handle a number of status–command and analogue inputs and outputs and a CRT (Vectran, n.d.).

---

* A partial list of manufacturers includes the following: Analog Devices, P.O. Box 280, Norwood, Massachusetts 02062; Analogic Corporation, Audubon Road, Wakefield, Massachusetts 01880.

† Vectran Corporation, 261 Kappa Drive, Pittsburgh, Pennsylvania 15230.

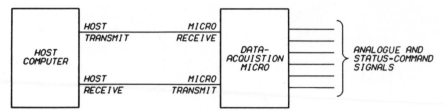

Fig. 4. Host computer to data-acquisition micro.

TABLE I

DECIMAL, BINARY, OCTAL, AND HEXADECIMAL EQUIVALENCE

Decimal	Binary	Octal	Hexadecimal
1	1	1	1
2	10	2	2
3	11	3	3
4	100	4	4
5	101	5	5
6	110	6	6
7	0111	7	7
8	1000	10	8
9	1001	11	9
10	1010	12	A
11	1011	13	B
12	1100	14	C
13	1101	15	D
14	1110	16	E
15	1111	17	F
16	10000	20	10
17	10001	21	11
18	10010	22	12
19	10011	23	13
20	10100	24	14
21	10101	25	15
22	10110	26	16
23	10111	27	17
24	11000	30	18
25	11001	31	19
26	11010	32	1A
27	11011	33	1B
28	11100	34	1C
29	11101	35	1D
30	11110	36	1E
31	11111	37	1F
32	100000	40	20

TABLE II

DECIMAL AND BCD EQUIVALENCE

Decimal	BCD	Decimal	BCD
0	0000 0000	14	0001 0100
1	0000 0001	15	0001 0101
2	0000 0010	16	0001 0110
3	0000 0011	17	0001 0111
4	0000 0100	18	0001 1000
5	0000 0101	19	0001 1001
6	0000 0110	20	0010 0000
7	0000 0111	25	0010 0101
8	0000 1000	30	0011 0000
9	0000 1001	35	0011 0101
10	0001 0000	40	0100 0000
11	0001 0001	45	0100 0101
12	0001 0010	50	0101 0000
13	0001 0011		

D. Digital Signal Representation

Binary numbers are used in the machine code actually executed by a computer. As the name implies, it is a base-two system with each binary digit, or bit, either on or off. Since a long string of ones and zeros is very cumbersome for people to handle, binary numbers may be represented with octal, hexadecimal, or decimal notations. Three digits (bits) may be represented by an octal (base-eight) digit ranging from 0 to 7. Four bits may be represented by a hexadecimal digit ranging from 0 to 15. To represent the values of 10 to 15, the letters A–F are used. Table I shows the comparisons among decimal, binary, octal, and hexadecimal notations.

Since humans have a habit of using decimal numbers (perhaps owing to the total of 10 digits on our hands), there must be a binary representation of the decimal system. The numbering system is called binary-coded decimal (BCD). In BCD, each decimal digit is represented by the binary values of 0 to 9; the letters A–F are not used. BCD is used for numeric displays, such as in LED and LCD displays. Table II shows decimal numbers with BCD equivalences.

Humans also use letters, punctuation, and other characters. In addition to the visible characters used by humans, computers and terminals need control characters. The most common system to represent the various characters is the American Standard Code for Information Interchange (ASCII) (Lear–Siegler, 1980); most printers and CRTs use the ASCII code (see Table III). Punched-card systems and some IBM terminals use the Extended Binary-Coded Decimal Interchange Code (EBCDIC).

E. Methods of Digital Signal Transmission

Digital signal transmission may be broken down into two types— . parallel and serial.

1. PARALLEL SIGNALS

In parallel signal transmission, each bit of a byte (group of eight bits) is assigned to a separate conductor. In addition, there may be control signals. All bits of a byte or larger group of bits are sent at the same time. The advantage to this method is speed; however, since each bit requires a separate, parallel conductor, parallel data transmission is not economical over long distances. Examples of parallel data transmission methods include CAMAC, IEEE-488, Centronics printer standard (also used by other printers), and the address and data busses inside computers (Leibson, 1983).

2. SERIAL SIGNALS

Serial methods of data transmission were developed long ago to enable data to be transmitted long distances over a pair of wires. The earliest practical example was the telegraph using Morse code. Before Morse, Andre Marie Ampere made a telegraph with 26 wires, one for each letter!

In serial transmission, bits of a character or byte are transmitted sequentially. There are two modes in serial transmission: asynchronous and synchronous. In synchronous transmission, characters (most often in ASCII form) or numbers are sent with no random time spacing between them. Sync characters are periodically transmitted to keep the receiver synchronized with the transmitters. In asynchronous transmission, each byte is synchronized individually for each character. A character is synchronized by a start bit, then data bits follow. Afterwards, a parity bit typically is transmitted to make the sum of the data bits and parity bit even (even parity) or odd (odd parity), followed by one or more stop bits. Unlike synchronous data transmission, characters or numbers may be sent individually at random times (Leibson, 1983).

TABLE III

ASCII Control-Code Chart[a]

B7 B6 B5 / BITS / B4 B3 B2 B1	CONTROL		NUMBERS SYMBOLS		UPPER CASE		LOWER CASE	
	0 0 0	0 0 1	0 1 0	0 1 1	1 0 0	1 0 1	1 1 0	1 1 1
0 0 0 0	NUL	DLE	SP	0	@	P	`	p
0 0 0 1	SOH	DC1	!	1	A	Q	a	q
0 0 1 0	STX	DC2	"	2	B	R	b	r
0 0 1 1	ETX	DC3	#	3	C	S	c	s
0 1 0 0	EOT	DC4	$	4	D	T	d	t
0 1 0 1	ENQ	NAK	%	5	E	U	e	u
0 1 1 0	ACK	SYN	&	6	F	V	f	v
0 1 1 1	BEL	ETB	'	7	G	W	g	w
1 0 0 0	BS	CAN	(	8	H	X	h	x
1 0 0 1	HT	EM	)	9	I	Y	i	y
1 0 1 0	LF	SUB	*	:	J	Z	j	z
1 0 1 1	VT	ESC	+	;	K	[	k	{
1 1 0 0	FF	FS	,	<	L	\	l	|
1 1 0 1	CR	GS	−	=	M	]	m	}
1 1 1 0	SO	RS	.	>	N	^	n	~
1 1 1 1	SI	US	/	?	O	_	o	RUBOUT (DEL)

LEGEND

octal 10 ←— LSI CURSOR CONTROL CODE
BS ASCII character
hex 8 8 decimal

[a] By permission of Lear–Siegler, Inc.

3. Primary Serial Methods

This section deals with serial methods used directly by computers and associated equipment. The most common method of serial digital transmission is RS-232C, originally developed for modems. In addition to signal receive, transmit, and ground, there are several different protocol lines that are frequently unused. The longest distance guaranteed by the RS-232C standard is 16.5 m (50 ft). RS-232C may be used for longer distances at the risk of the user (Leibson, 1980). To increase distances and provide more flexibility, the RS-449 standard was developed.

An older method, and one good for long distance, is the 20-mA digital current loop originally developed for Teletype terminals. The current-loop method may be used for several hundred meters.

4. Secondary Serial Methods

Since the above methods are limited in distance and one would like to communicate over many kilometers, line drivers, modems, and fiber optics were developed.

Line drivers may be used in a plant with dedicated lines. An example is shown in Fig. 5. Line drivers may be used to transmit data at 9600 baud (bits per second) for distances of several kilometers.

The name "modem" is an abbreviation of MODulate–DEModulate. Modems were developed to transmit digital signals over ordinary telephone lines and are not limited in distance. If verbal communication can be accomplished over the lines, then the lines are adequate to transmit data by modems. Modem transmission rates are typically much slower than maximum line-driver transmission rates. Most modems run at only 300 baud; some transmit at 1200 baud. There are special modems capable of transmitting at 9600 baud for unlimited distances. A typical modem transmits two different frequencies, one for the high state of the RS-232C signal and one for the low state (Gandalf Technologies, 1982).

An ideal method of transmitting data in an electrically noisy, potentially explosive industrial environment is fiber optics, since the fibers are not electrically conductive. Fiber optics can transmit data of high speed; 40 Mbits/sec (megabaud) is possible. At lower speeds, the maximum distance between the transmitter and receiver may be 5 km. Unfortunately, difficulty in establishing connections in the field has retarded application in industry; as more experience in

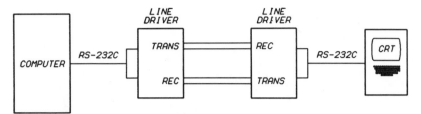

Fig. 5.   Line-driver diagram.

making splices and connections is gained, fiber optics will be used increasingly (Persun, 1982).*

5. ERROR CHECKING

In digital data transmission there is a possibility that one or more bits of a digital number or character may be erroneous. A common error-checking method uses parity as follows: An extra bit (such as an ASCII character) added to a binary word is set on or off to achieve even or odd parity. If a bit in the binary word is dropped or erroneously high, then the total number of bits will be an odd number for even parity or an even number for odd parity; an error condition then exists, and the byte may be rejected. However, the method is not foolproof; more than one bit may be dropped or erroneously high (Leibson, 1983).

In those applications where error checking is especially crucial, the checksum method is used, where a sum (or the least-significant portion of the sum) of the transmitted block of characters is transmitted. The receiving device then sums the characters and compares the sum with the received checksum. If the two do not agree, then the entire transmission block is rejected (Analog Devices, 1981).

F. Programming

1. MACHINE AND ASSEMBLY LANGUAGES

Machine language is the set of binary number instructions that is actually used by the computer to execute programs. Since binary numbers are cumbersome for human programmers to use, the binary

* Line drivers, modems, and fiber-optic equipment are available from Gandalf Data, Inc., 1019 South Noel, Wheeling, Illinois 60090.

```
*******************************************************
*      M O D C O M P   I I   A S S E M B L Y   L A N G U A G E
*******************************************************
*    THIS PROGRAM DEMONSTRATES SOME OPERATIONS WITH ASSEMBLY LANGUAGE.
*    LINES STARTING WITH ASTERISKS ARE COMMENT LINES.
*
*    THE LINE BELOW CONTAINS DESCRIPTIONS OF THE DIFFERENT FIELDS.
*
*LBL  INSTRUCTION  ARGUMENT  COMMENTS
*1    2            3         4
*
```

Line	Address	Hex	LBL	INSTRUCTION	ARGUMENT	COMMENTS
12				PGM	DMSTR	NAME THE PROGRAM "DMSTR".
13	0000	ED10		LD1,1	10	LOAD THE VALUE OF 10 INTO REGISTER 1.
	0001 A	000A				
14	0002	E530		LDM,3	EXCL	LOAD INTO REGISTER 3 THE CONTENTS
	0003 R	0039				STORED AT LOCATION (LABEL) "EXCL".
15			*			
16	0004	E540		LDM,4	ZERO	LOAD INTO REG. 4 THE VALUE LOCATED AT "ZERO".
	0005 R	0038				
17	0006	E840		ADI,4	#30	ADD HEXADECIMAL 30 TO REG. 4 GIVING
	0007 A	0030				ASCII FORM OF INTEGER.
18			*			
19	0008	2038		LLS,3,8		SHIFT CONTENTS OF REGISTER 3 EIGHT TIMES.
20	0009	4843		ADR,4,3		ADD CONTENTS OF REG 3 TO REG 4.
21			*			AT THIS POINT, REGISTER 4 CONTAINS ASCII CHARACTERS FOR EXCLAMATION
22			*			AND THE NUMBER 0, OR HEXADECIMAL 2130.
23	000A	E640		STM,4	ASCNR	STORE CONTENTS OF REGISTER 4 AT
	000B R	0036				LOCATION "ASCNR".
24			*			
25			*			WRITE CHARACTERS, STARTING AT LOCATION "MESG".
26			*			FIRST OF ALL, THE LOCATION OF THE USER'S FILE TABLE MUST BE LOADED INTO
27			*			REGISTER 2. THE WRITE USER FILE TABLE CONTAINS PARAMETERS NECESSARY FOR
28			*			THE WRITE.
29			*			
31	000C	ED20		LD1,2	WRUFT	LOAD LOCATION OF WRITE USER FILE TABLE
	000D R	0017				INTO REGISTER 2.
32			*			
33	000E	2301	AGAIN	REX,#1		PERFORM WRITE (REQUEST EXECUTIVE HEX 1).
34	000F R	001D		DFC	MESG	STARTING LOCATION FOR MESSAGE.
35	0010 A	0034		DFC	52	NUMBER OF BYTES TO BE WRITTEN.
36	0011	80F0		ABM1,15	ASCNR	ADD 1 (BIT 15) TO THE EXCLAIMATION,
	0012 R	0036				
37	0013	711F		SBRB,1,15	AGAIN	SUBTRACT 1 (BIT 15) FROM REGISTER 1 AND
38	0014 R	000E				ASCII NUMBER.
39			*			BRANCH BACK TO LOCATION "AGAIN" IF CONTENT
40			*			OF REGISTER 1 IS NOT ZERO.
41	0015	2307		REX,#7		WRITE END-OF-FILE FOR DEVICE 6.
42	0016	2312		REX,#12		TERMINATE PROGRAM.
43	0017 A	0000	WRUFT	DFC	0,#6,#A000,0,0,0	WRITE USER FILE TABLE
	0018 A	0006				
	0019 A	A000				
	001A A	0000				
	001B A	0000				
	001C A	0000				

Fig. 6. MODCOMP II assembly program example. (By permission of Modular Computer Systems, Inc.)

```
44
45    *
      MESG  DFC    "0 THE PRINTER IS DISPLAYING EXCLAMATION "
001D A 3020
001E A 5448
001F A 4520
0020 A 5052
0021 A 494E
0022 A 5445
0023 A 5220
0024 A 4953
0025 A 2044
0026 A 4953
0027 A 504C
0028 A 4159
0029 A 494E
002A A 4720
002B A 4558
002C A 434C
002D A 4149
002E A 4D41
002F A 5449
0030 A 4F4E
0031 A 2020
46          DFC    "INTEGER "
0032 A 494E
0033 A 5445
0034 A 4745
0035 A 5220
0036 A 0000
47    ABDNR DFC    0       ARGUMENT INITIALLY SET TO ZERO, BUT IT WILL
48    *                    CONTAIN ASCII CHARACTER ! AND AN ASCII
49    *                    INTEGER.
50    0037 A 0000  DFC  0  NULL TERMINATOR
51    *                    FORM OF A SINGLE-DIGIT INTEGER.
52    ZERO  DFC    0       INTEGER 0 IS BEING STORED AT LOCATION
      0038 A 0000          LABEL "ZERO".
53    EXCL  DFC    #21     ASCII EXCLAMATION CHARACTER
0039 A 0021
54    003A   END
55

THE PRINTER IS DISPLAYING EXCLAMATION   INTEGER !0
THE PRINTER IS DISPLAYING EXCLAMATION   INTEGER !1
THE PRINTER IS DISPLAYING EXCLAMATION   INTEGER !2
THE PRINTER IS DISPLAYING EXCLAMATION   INTEGER !3
THE PRINTER IS DISPLAYING EXCLAMATION   INTEGER !4
THE PRINTER IS DISPLAYING EXCLAMATION   INTEGER !5
THE PRINTER IS DISPLAYING EXCLAMATION   INTEGER !6
THE PRINTER IS DISPLAYING EXCLAMATION   INTEGER !7
THE PRINTER IS DISPLAYING EXCLAMATION   INTEGER !8
THE PRINTER IS DISPLAYING EXCLAMATION   INTEGER !9
```

numbers are commonly represented by octal numbers (each digit representing three bits) or hexadecimal numbers (each digit representing four binary bits) (see Section III,D).

Since the process of programming in machine code is tedious and prone to error, a more convenient method of programming was devised, known as assembly language. Figure 6 shows an example of assembly-language programming for a MODCOMP (Modular Computer Systems, Inc.) computer. To the right of the two machine-language columns is the assembly code that was entered into the computer by the programmer, using punched cards or CRT. The first character field is the symbolic-address–label column, labeled LBL. The next character field is the opcode, or the instruction field. The third character field is the operand, or the argument of the instruction or op code. Note that in several cases in the example the argument refers to the symbolic-address label elsewhere in the program. Finally, the rightmost character field contains comments. If more room is needed for comments, then entire lines may be set aside by putting an asterisk in the first position of the first character field (MODCOMP convention).

The code represented by two number fields, each four digits long, on the left-hand side of the listing is the machine code used by the computer. The left-hand column contains the relative addresses of the machine code. The symbolic-address labels refer to certain locations in the machine addresses. The right-hand column contains the machine instructions, which correspond to the opcodes (instructions) and operands (arguments) in the assembly instruction parts.

The entity that generates the machine code from the assembly language is an assembler, which is actually a special program. As mentioned before, the assembly program, or source code, written by the programmer may be entered by punched card or CRT. The assembler then produces the machine code, or object code. Figure 7 shows the steps in preparing the machine code from assembly languages.

Since assembly language is a close representation of the machine code, assembly language for one computer will not work for another computer unless the assembly language for the first computer is a subset of or the same as the assembly language for the second computer. Also, assembly language is cumbersome for routine use. Hence, higher-level languages have been developed. However, assembly language programming is still used for applications requiring extensive bit manipulations, special input–output operations, and certain real-time applications.

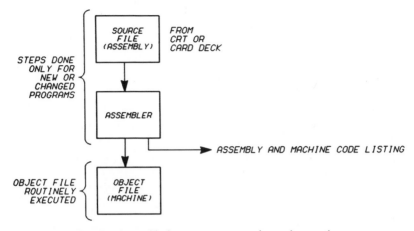

Fig. 7.   Assembly-language programming and execution.

2.   HIGH-LEVEL LANGUAGES

To program in machine and assembly languages, the programmer must know the internal architecture of the computer. Since there is no universal standard for computer architecture, the programmer must be familiar with each different type of computer. Also, programs written for one computer are not likely to execute on another computer type. Hence, high-level languages have been developed, examples of which include FORTRAN, BASIC, Pascal, PL1, ADA, and COBOL. A high-level program developed for one computer can be successfully entered into and executed on a second computer, even if the second computer has a different architecture, as long as the language standards are the same for both computers. The concept is known as *portability*. Since programming (software) costs greatly exceed hardware costs, program portability enables costs to be distributed over more computer installations.

High-level languages may be classified as compiled or interpretive.

(i)   Compiled languages are those in which the object file (machine code) is created or compiled from the source file (the characters from the programmer, typically entered through a CRT or card reader) only when there are new programs or when old programs are changed. Figure 8 diagrams the procedure for producing the executable object file (machine code) from the source file. With compiled languages, only the object file is routinely executed. In some com-

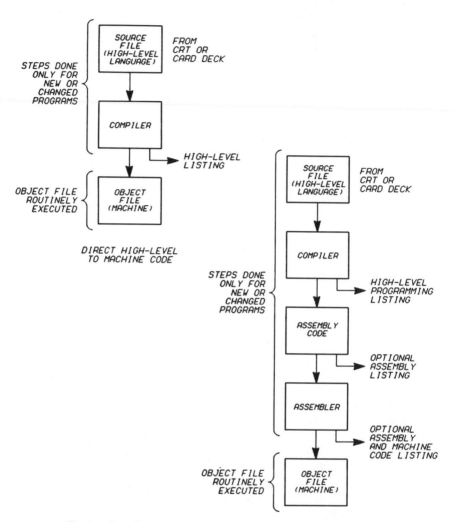

Fig. 8. Compiled high-level–language programming and execution.

puter systems, the object (machine) code is created directly from the source code (from the programmer). In other systems, the high-level–language program is compiled to assembly language. The assembly-language file is then assembled to machine code. The advantage of the second method is that the program may be manually optimized at the assembly stage by the programmer. Most high-level languages except interpretive BASIC and APL are compiled languages. Figure 9 shows an example of a FORTRAN program.

```
 1              PROGRAM DEMSTR
 2      C          THIS IS A FORTRAN PROGRAM TO DO MUCH THE SAME THING AS THE
 3      C          ASSEMBLY PROGRAM DMSTR.
 4      C          SET I TO 0:
 5              I = 0
 6      C          SET THE COUNT VARIABLE TO 10.
 7              ICOUNT = 10
 8      C          CONVERT INTEGER I (SET TO 0) TO ASCII BY ADDING HEX 30:
 9              IASCI = I + 2Z30
10      C          SET IEXC TO THE ASCII EXCLAIMATION POINT:
11              IEXC = 2Z21
12      C          SHIFT IEXC LEFT BY 8 BITS (MULTIPLY BY HEX 100 OR DECIMAL 256):
13              IEXC = IEXC * 3Z100
14      C          ADD IEXC TO IASCI TO FORM TWO ASCII CHARACTERS "! 0"
15              IASCI = IASCI + IEXC
16      C          WRITE TO THE PRINTER THE ASCII AND HEXADECIMAL FORMS
17      C          OF IASCI.
18        30 WRITE(6,32) IASCI,IASCI
19        32 FORMAT('0EXCLAMATION MARK WITH NUMBER IS ',A2,
20           * ' AND THE HEXADECIMAL VALUE IS ',Z4)
21      C          DECREMENT COUNT VARIABLE:
22              ICOUNT = ICOUNT - 1
23      C          ADD 1 TO IASCI:
24              IASCI = IASCI + 1
25              IF(ICOUNT.GT.0) GO TO 30
26      C          WRITE END-OF-FILE FOR DEVICE 6.
27              ENDFILE 6
28              CALL EXIT
29              END
MULT
EDIT    MAIN BI
WEO     BO
EXIT

EXCLAMATION MARK WITH NUMBER IS !0 AND THE HEXADECIMAL VALUE IS 2130

EXCLAMATION MARK WITH NUMBER IS !1 AND THE HEXADECIMAL VALUE IS 2131

EXCLAMATION MARK WITH NUMBER IS !2 AND THE HEXADECIMAL VALUE IS 2132

EXCLAMATION MARK WITH NUMBER IS !3 AND THE HEXADECIMAL VALUE IS 2133

EXCLAMATION MARK WITH NUMBER IS !4 AND THE HEXADECIMAL VALUE IS 2134

EXCLAMATION MARK WITH NUMBER IS !5 AND THE HEXADECIMAL VALUE IS 2135

EXCLAMATION MARK WITH NUMBER IS !6 AND THE HEXADECIMAL VALUE IS 2136

EXCLAMATION MARK WITH NUMBER IS !7 AND THE HEXADECIMAL VALUE IS 2137

EXCLAMATION MARK WITH NUMBER IS !8 AND THE HEXADECIMAL VALUE IS 2138

EXCLAMATION MARK WITH NUMBER IS !9 AND THE HEXADECIMAL VALUE IS 2139
```

Fig. 9.   MODCOMP II FORTRAN program example. (By permission of Modular Computer Systems, Inc.)

(ii) Interpretive systems store the program as alphanumeric characters, typically ASCII. Interpretive BASIC (distinguished from compiled BASIC) is an example. The computer converts the alphanumeric characters to machine code as the program is executed. Unlike compiled languages, the conversion or interpretation is done each time the program is executed, as shown in Fig. 10. Interpretive systems are slower than compiled systems; compiled programs do not need conversion from characters to machine code for routine program execution. However, interpretive languages are simpler to use; the program may be modified without compilation,

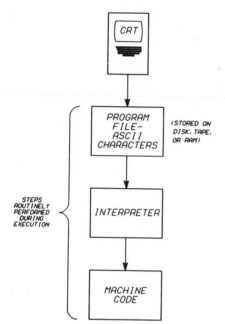

Fig. 10. Interpretive-language programming and execution.

and error messages will appear, if required, after each line is entered. Hence, debugging an interpretive program is simpler than a compiled program, since error messages for compiled programs show up only when the entire program is compiled.

To obtain the best from both systems, a program may be developed and executed with interpretive BASIC on a trial basis. If the program execution is too slow, the program may be compiled with a BASIC compiler (Coats, 1982).

G.   Computer Operating Systems

1.   DEDICATED COMPUTER SYSTEM

The dedicated computer is a small computer with only one task or program to execute in its operation. Most dedicated computers are microprocessor based. Applications include consumer appliances (microwave ovens, washers, dryers, automobiles, etc.) and instruments (test instruments, single-board data-acquisition microcomputers, data loggers, etc.)

2. TIME-SHARING SYSTEMS

A time-sharing system enables several users to use the system at the same time. Each user may be assigned a separate CRT. In time sharing, a scheduler in the operating system coordinates the execution of various programs. As resources permit, the scheduler brings in programs (object files) from a bulk-storage device (disk or tape) into the computer's random access memory (RAM) (core or solid state) and initiates execution. Different programs may be assigned different priorities. If a low-priority program is being executed and a higher priority program is requested, the scheduler will suspend operation of the low-priority program and initiate the higher-priority program. After the higher-priority program is complete, the scheduler will resume execution of the low-priority program. If two programs have the same priority, then the scheduler may rotate execution and suspension between the two programs. A special case of a time-sharing system is a real-time system, where programs may be executed only at certain times. In a real-time system, the scheduler uses a system timer. Also, delays may be put in programs to enable the computer to control external equipment or to communicate with slow devices (Freeman and Perry, 1977).

H. Computer Networks

In a large plant such as a refinery, data processing and control may be performed with a network of computers, instead of with only a single computer. With a computer network, only one part of the plant may be affected if one computer goes down. At the author's plant, most computers are MODCOMP (Modular Computer Systems, Inc.) computers, which are physically connected together by dedicated phone lines. In this system, data may be retrieved from and written into remote computer disk files as simply as if the disk files were connected to the local computer. However, data access is slower since the phone lines (driven by line drivers) are limited to 9600 bits/sec. Computers of other makes also communicate with some of the MODCOMP computers. However, each software interface between two different brands of computers has to be created on an individual basis, and data transfer is much slower and more prone to error than between two MODCOMP computers. Figure 11 shows a partial diagram of a refinery system's computer network.

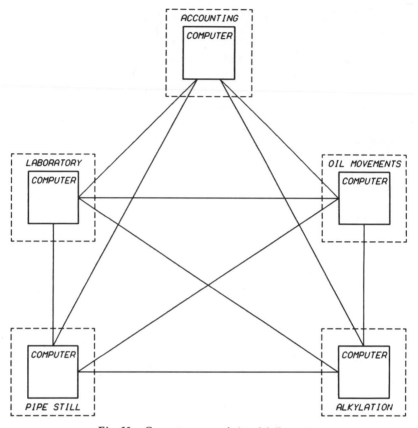

Fig. 11.   Computer network (partial diagram).

IV.   Conclusion

Process analyzer data can be displayed on a computer terminal (CRT or printer) provided that several things have been done. First, the process analyzer must be connected by some link to a computer input. Second, the process analyzer must send the appropriate signals down the link. Third, computer programs must be available to collect the data from the computer input and store them on bulk memory (disk or type). Finally, there must be one or more programs to retrieve the data from the bulk memory, convert them to a suitable format, and display them on the terminal.

References

Analog Devices, Inc. (1981). *User's Manual, μMac 4000, μMac 4010.* Norwood, Massachusetts.

Andreiev, N. (1982). *Control Eng.* **29**, 82–85.

Andrew, W. G., and Williams, H. B. (1979). "Applied Instrumentation in the Process Industries," Vol. 1, pp. 46–195, 209–249. Gulf Publishing, Houston, Texas.

Coats, R. B. (1982). "Software Engineering for Small Computers," pp. 65–73. Reston Publishing Co., Inc., Reston, Virginia.

Conlog, Inc. (1980). *Isoterm (TWT × LPI) Loop Powered Isolator* sales brochure. Elk Grove Village, Illinois.

Freeman, D. E., and Perry, O. R. (1977). "I/O Design—Data Management in Operating Systems," pp. 1–23. Hayden, Rochelle Park, New Jersey.

Gandalf Technologies, Inc. (1982). Gandalf Data Communications Products. Wheeling, Illinois.

Lear–Siegler, Inc. (1980). "ADM-31 Data Display Terminal Operators Quick Reference Guide," p. A1. Anaheim, California.

Leibson, S. (1980). *Instruments Control Syst.* **53**, 47–53.

Leibson, S. (1983). "The Handbook of Microcomputer Interfacing," pp. 90–158. Tab Books, Blue Ridge Summit, Pennsylvania.

Liptak, B. G. (1970). "Instrument Engineers' Handbook, Volume II, Process Control," pp. 615–661, 873–977. Chilton, Philadelphia.

Persun, T. (1982). *Instruments Control Syst.* **55**, 50.

Soisson, H. E. (1975). "Instrumentation in Industry." Wiley, New York.

Vectran Corporation (n.d.). VR-20, VR-30, and VR11 sales brochures. Pittsburgh.

15

Process Analyzer Optimization in Control Systems

LEONARD J. WACHEL*

Process Analyzer Group
Amoco Oil Company
Whiting, Indiana

I. Introduction: Analyzers as Analytical Inputs

The characterization of unit feed and product streams in terms of physical properties or chemical compositions is an essential element in the evaluation of process performance. This characterization requires physical and chemical analyses utilizing a wide range of analytical instrumentation and sampling techniques. Certain characterizations are suitable for on-stream process analysis while others are

* Present address: UIC, Incorporated, Joliet, Illinois 60434.

best handled off stream by unit operators or laboratory technicians. This chapter will discuss primarily the optimization of on-stream process-analyzer installation and performance to levels suitable for control applications. Off-stream analysis will be discussed only briefly as it applies to collecting process-stream data and providing for analyzer validation.

II. Process Analysis for Process Control

A. On-stream Analysis

Many process plants in operation today are primarily controlled by the monitoring of pressure, temperature, flow, and liquid level. These measurements are direct and reliable, and associated instrumentation is relatively easy to install, calibrate, and maintain. Engineering and maintenance personnel are often reluctant to install and use process analyzers because of their reputation for low reliability and high maintenance requirements. Instead, some individuals prefer to maintain control by using easily measured physical properties and intermittent laboratory analyses as a limited and possibly nonrepresentative data-base from which to make adjustments of controller set points or to target values for process variables.

Frequently, process analyzers are instruments with which the engineer has only a passing familiarity, having worked with similar basic units in laboratory courses or having studied and operated larger-scale pilot-plant or production units. Although operating principles and basic elements of design between process analyzers and other units may be similar, many significant differences exist. The special construction of a process analyzer, which is required to operate safely and dependably in a hazardous, corrosive environment, is often not well understood by the manufacturer's representative or even the engineer who is to specify and apply an analyzer to a particular process.

The following sections will explore these considerations in depth, to explain how they should be addressed and to familiarize the reader with the mass of reference materials available. For many individuals, the skills required for analyzer specification and application are only occasionally needed and often never developed, yet access to this knowledge has become increasingly important with increases in energy and labor costs and the decline of operating profit margins.

On-stream analysis employs process analyzers that continuously and automatically sample, measure, and report the physical properties or chemical compositions of monitored process streams. Off-stream analysis can provide the same types of determinations as process analyzers but typically require manual sampling for later analysis in facilities that are frequently remotely located with respect to the process unit.

Table I shows a representative sampling of the types of determinations frequently made using process instrumentation.

B. Advantages of Using Process Analyzers

Physical-property or composition analyzers are used to "fine tune" a process operation to optimal levels not otherwise possible. Formerly, results were obtained by periodically submitting process grab samples to the laboratory for analysis. Typically, the lab results obtained some hours later were used in an attempt to control the process. The dynamic nature of the process and the long time lags in both the analysis and reporting of results raise serious questions about how well the data represent ongoing process conditions. At times, process adjustments made as a result of laboratory analyses actually move the process farther away from the desired optimal conditions. For these reasons, analyzers connected directly to the process are needed to provide instantaneous or recent data results, typically within minutes or less. Other benefits include the elimination of sampling and labeling errors, the detection of transient process changes, and a considerable reduction in the cost per analysis. All of these benefits can be achieved with well-selected analyzers having high performance, good accuracy, and low maintenance requirements. In addition to a high-quality, well-protected analyzer, a well-designed sample system that performs continuously with short transport times and a convenient means of calibration are also essential elements of an optimum analyzer system.

III. Sampling Systems for Process Analyzers

Most analyzers are designed to work dependably when provided with a clean, dry, noncorrosive, homogeneous sample that is maintained at regulated temperature, pressure, and flow conditions. Process samples that fail to meet analyzer requirements adversely, and sometimes fatally, affect analyzer performance. Even the best analyzers on the market cannot provide reliable results if their associ-

TABLE I

TYPICAL PROCESS INSTRUMENT DETERMINATIONS

Analyzer	Principle	Determination
Physical Properties		
BTU	Calorimetry or gas chromatography	Heat capacities of gases
Cloud point	Light variation detected by a photocell	Crystal formation
Density	Ultrasonic or mass separation	Specific gravity or density of liquid or gas
Distillation–boiling point	Fractionation or chromatography	Percentage recovery or temperature
Flash point	Heat rise using thermocouple or RTD	Flash temperature of flammable material
Octane	Heat of combustion using thermocouple	Correlated octane rating
Pour point	Resistance to flow via D–P cell or moving sensor	Pour point of material
Reid vapor pressure	Equilibrium pressure measured at elevated temperature or by gas chromatography	Vapor pressure direct or by correlation
Viscosity	Ultrasonic, shear, or capillary tube	Resistance to flow
Chemical Properties		
Ambient air	Gas chromatograph, catalytic cell, spectrographic analysis	Concentration of flammable or toxic gases in air
Combustible	Catalytic cell	Combustible gases, lower explosive limit (LEL)
Conductivity	Ion transfer	Liquid purity, total dissolved solids (TDS)
Gas analyzer (SO_2, CO_2, etc.)	UV/visible/IR spectral absorption	Gas composition, concentration
Gas chromatograph	Gas–liquid phase distribution on a separate column	Gas or liquid composition
Hydrogen sulfide	Chemical reaction	H_2S in liquid or gas
Moisture	Adsorption, capacitance, electrolysis	Water in gas or liquid
Oxygen	Paramagnetic, polarographic, zirconium oxide fuel cell	Oxygen in gas or liquid
Opacity	Spectral absorption	Visible stack emissions
pH	Ion transfer	Water quality
Sulfur	Radiation absorption–emission	Percentage sulfur in fuels

ated sample systems do not deliver a representative process sample that meets the analyzer's requirements. The sample system of a process analyzer is often its weakest link and the principal cause of lower-than-projected levels of performance and reduced economic benefit.

The following subsections will define the primary objectives of sample systems, discuss system design, and recommend reliable hardware so the reader can select or design a sample system that will not hinder optimum analyzer performance.

A. Sample-System Objectives

Too often, vendors of on-stream process analyzers concentrate on the features and the sale of an analyzer, leaving the design of the sampling system to the user. Most vendors and users recognize that the majority of sampling requirements involve much more than a sample tube connecting the process and the analyzer. However, varying levels of vendor assistance in sample-system design are available when requested. Materially, sample systems consist of filters, coalescers, pressure regulators, vaporizers, coolers, gauges, valves, flowmeters, pumps, traps, sample lines, steam and electric tracing, multiple utilities, and protective and mounting hardware. A vendor frequently does not have complete details of the application objectives or design of the process unit and as a result cannot make satisfactory, detailed recommendations to the user.

Each installation is to a varying degree unique and requires individual review. Such review is essential in providing data for the determination of sample takeoff and return points and analyzer placement. The payoff will be a higher operating factor and reduced levels of maintenance.

The sample system must extract a representative sample from the process, maintain the integrity of the sample during transport to the analyzer, properly condition the sample for the analyzer, and then return the analyzed and excess sample to the process or properly dispose of it. Each of these requirements will be discussed in greater detail.

1. EXTRACTING A REPRESENTATIVE PROCESS SAMPLE

The challenge is to obtain a representative sample of the components to be measured while leaving behind undissolved water, particulates, or other undesirable materials that are typically present in the process stream. The selection of the sampling location is of critical importance. Process-stream locations with conditions of high turbulence, high levels of particulates, mixed phases of vapor and liquid, entrained contaminants, and high temperatures and pressures should be avoided since it may be extremely difficult if not impossible to obtain a representative sample. Even the best-designed sampling and analyzer systems are doomed to failure unless they are provided with a representative sample of the process at the sampling point. The following guidelines are useful in the process of sample-point selection.

Sample the process stream at a point where the process variables to be measured are all present as a single phase, liquid or gas. Mixed-phase streams are much more difficult to sample and should be avoided whenever possible. The actual sample point should be located in a vertical section of process piping, at least five pipe diameters away from any bends, pumps, valves, or restrictions that may cause turbulence. If it is necessary to sample from a horizontal section of piping, always sample from the side, never from the top or bottom. Undissolved gases and vapors travel in the top of horizontal lines, and sediments, corrosives, undissolved water, etc., travel along the bottom. Sampling at these less-desirable points can adversely affect sample quality and increase the incidence of repairs of both the process analyzer and its sampling system.

Further improvements in sample quality can be obtained by the use of a sample probe. A typical sample probe screws directly into a 1-in. process valve that has been welded to the larger-diameter process pipe at the desired sample point. The probe itself consists of a length of rigid instrument tubing, a Conax packing gland, a ferrule, and an instrument shut-off valve. The diameter of the sample-probe tubing must be small enough to allow passage through the 1-in. process valve and not smaller than the diameter of the tubing that will transport the process sample to the analyzer. Typical tubing sizes vary between $\frac{1}{4}$ and $\frac{1}{2}$ in. for both gas- and liquid-sampling applications. The length of the sample-probe tubing is adjusted to allow passage through the valve to the center of the pipe. As pipe diameter increases, it becomes less important to sample the center of the stream. As a rule, for piping greater than 1 ft in diameter, allow the probe to extend into the pipe for a distance of one-third the inside diameter of the process pipe.

The Conax packing gland screws into the 1-in. process valve and allows the insertion and retraction of the sample-probe tubing into a process stream during normal operation. Once cut to the desired length, the sample-probe tubing is equipped with a ferrule at the process end, the Conax gland and attaching hardware are placed into approximate position, and an instrument shut-off valve is permanently secured to the analyzer end of the sample-probe tubing. The ferrule acts as a safety device, preventing full retraction of the sample probe through the Conax gland by maintenance personnel or high process pressures. The sample probe is inserted by securing the threaded portion of the Conax gland to the process valve. The instrument valve should be closed. The seating portion of the Conax valve should be tightened until the probe tubing can be slid forward only with difficulty. The probe is then slid forward until it touches the gate of the process valve. The process valve is slowly opened, and the probe tubing is slid forward against stream pressure into its final position. There should be no leaks at this point with the exception of minimal leakage around the Conax gland. Once the probe is positioned, the seating portion of the Conax gland is tightened, and there should be no detectable leakage from the probe assembly. If it becomes necessary to remove the sample probe, it is imperative that the probe be retracted past the gate of the process valve; otherwise, the probe will be crushed when the process valve is closed, making it impossible to seal off the process stream.

2. MAINTAINING SAMPLE INTEGRITY DURING TRANSPORT

Sample transport distances are kept to a minimum by locating the analyzer as close as possible to the sampling point, minimizing sample pressure drops and transport times. The only trade-offs occur in the consideration of multiple analyzers sharing a common location, analyzers with multiple sample takeoff points, and the accessibility of the analyzer for calibration and maintenance.

Process samples should be transported from the process stream to the analyzer location either entirely as liquid or entirely as gas. Suppose that a process stream consists of a gaseous mixture of hydrocarbons at an elevated temperature and pressure and that a sample stream of this material is provided to a process analyzer. If this sample stream is allowed to cool, the dew point of the sample may be reached. At this dew point, the gaseous materials with the highest boiling points begin to condense as liquid droplets in the sample tubing. The result is an instantaneous change in the composition of the sample stream; it no longer represents the composition of the process stream at the time sampling occurred. In turn, the analyzer reports results based on a gaseous mixture devoid of the higher-boiling gases present in the process stream. Suppose further that the analyzer itself operates at a sufficiently high temperature to revaporize the liquid droplets as they reach the analyzer. As each liquid droplet is vaporized, a gaseous volume approximately 200 times that of the liquid droplet is generated. The analyzer would then report results based primarily on the analysis of a mixture of high-boiling gases. Under these conditions no meaningful data can be reported by the process analyzer even if it is in perfect working order. Any results that match the instantaneous composition of the process stream would be purely coincidental. For analogous reasons, it is equally important to maintain liquid samples as entirely liquid (Houser, 1972).

3. SAMPLE-LINE SIZING

It is important to calculate the sample transport time from the sample takeoff point to the analyzer for every analyzer installation; for analyzers in control applications, such calculations are essential. The analysis of a process stream whose composition can change quickly must therefore be performed quickly. Both the sample transport time to the analyzer and the analysis cycle time must be kept to minimum values.

The sample transport time is a function of the volume of all tubing and components between the sample takeoff point and the analyzer and the sample flow rate as shown below:

$$\text{transport time (min)} = \frac{\text{volume of all transport components (cm}^3)}{\text{sample flow rate (cm}^3/\text{min})}.$$

Notice that the smaller the size of the transport tubing, the shorter the transport time, as long as there are no significant reductions in flow rate. However, for liquid samples, the sample flow rate is reduced significantly due to pressure drop along the length of the sample line. The smaller the diameter of the sample tubing, the greater the pressure drop per unit length. Unless sample takeoff pressures are sufficiently high to allow for a considerable pressure drop and reduction in flow rate, it bcomes necessary to increase the diameter of the sample transport tubing. The most commonly used size of liquid-sample tubing is $\frac{1}{2}$ in. which takes into consideration all of the variables discussed previously.

A common liquid-sample analyzer installation will be described by way of illustration. The analyzer-sample takeoff and return points are selected to be the discharge and suction sides of a process pump. A ½-in. stainless steel tube is used to transport analyzer sample to the vicinity of the analyzer via a continuous-flow sample loop. A tubing tee is installed in the sample loop as close to the analyzer as possible. A short length of tubing connects the sample loop and the analyzer, maintaining the integrity of the sample, near the analyzer. The bulk of the analyzer sample is unprocessed and returned directly to the process stream. Only a small amount of sample is removed through the sample tee, reduced in pressure, filtered, flow-regulated, and analyzed. Ideally, all of the sample withdrawn from the process should be returned to the process, typically at a point of lower pressure. The discharge of analyzer samples into hydrocarbon drains or plant gas lines should be avoided whenever possible for both environmental and economic reasons.

The sample transport time is equally important to gas-sample analyzer installations. To optimize the trade-off between sample transport time and pressure drop, a pressure regulator is installed after the instument valve of the sample probe and set to about 10 psig. Reductions in flow rate due to pressure drops across the length of sample tubing are much less for gases at low pressure. Also, most analyzers are pressure sensitive and operate at atmospheric pressure. Typically, analyzers are provided with a process sample at only a slightly elevated pressure, and the sample to be analyzed is referenced to atmospheric pressure just before analysis. The small variations in atmospheric pressure on a day-to-day basis introduce negligible error in most cases. In consideration of the factors outlined previously, the most commonly used transport tubing for gas samples is ¼-in. stainless steel tubing, with a flow rate of 100 to 1500 cm^3/min.

Locating the pressure regulator near the analyzer for a gas-sample analyzer installation frequently results in the sample transport time being unacceptably long for the purpose of process analysis. This problem may be solved by using a sample loop of larger-diameter tubing as in the liquid-sample illustration and reducing the pressure of a small slip stream in the vicinity of the process analyzer.

4. DEW POINTS AND BUBBLE POINTS OF HYDROCARBON MIXTURES

To ensure that a gaseous sample reaching a process analyzer has exactly the same composition as the process stream at the instant of sampling, it is essential that the temperature and pressure be controlled such that the sample remains entirely gaseous. In addition to sample integrity, dual phases interfere with the operation of sample-system components such as pressure regulators, flow meters, and needle valves, further reducing analyzer performance.

If there is only one condensable component in a gas stream, its dew point at a particular temperature and pressure can be readily obtained by observing the intersection of the pressure and temperature lines on the dew-point curve of the component in question. Tables for dew-point and bubble-point calculations can be found in most gas-data books. The bubble point of a liquid mixture with only one volatile component is obtained in a similar fashion by noting the intersection of the pressure and temperature curves on the bubble-point curve of the component in question.

If a typical gas mixture contains several condensable components, each component contributes to the dew point of the mixture. The calculation becomes an iterative procedure where one arbitrarily selects a dew-point temperature and, using a vapor–liquid equilibrium chart, finds the corresponding pressure where M = 1 using the formula

$$M_1/K_1 + M_2/K_2 + M_3/K_3 + \cdots + M_n/K_n = M,$$

where M_1, M_2, M_3, ... M_n are the mole fractions of the components of the hydrocarbon mixture, K_1, K_2, K_3, ... K_n the vapor–liquid equilibrium constants at the dew-point temperature selected, and M is the total moles. If the total moles M is found to be greater than one for a fixed temperature, then the dew-point pressure is lower than the pressure indicated. If M is found to be less than one for the same fixed temperature (at a different pressure), then the actual dew-point pressure can be found by interpolation between the two values.

Similarly, the bubble point of a hydrocarbon mixture is due to the contributions of several volatile components. The calculation is also an iterative procedure where one finds a temperature at a fixed pressure (or a pressure at a fixed temperature) where M = 1, using the formula

$$M_1 K_1 + M_2 K_2 + M_3 K_3 + \cdots + M_n K_n = M,$$

where the M_n and K_n values are as previously defined, except that the K_n values are for bubble-point temperature.

Once a dew point is determined for a particular gaseous process stream, it is desirable to maintain the sample stream at a temperature 50°F above the dew-point temperature or at a pressure 20 psi below the corresponding dew-point pressure to ensure that the sample remains a gas. In a similar fashion, once a bubble point is determined for a particular liquid process stream, it is desirable to maintain the sample stream at a temperature 50°F below the bubble-point temperature or at a pressure 20 psi above the corresponding bubble-point pressure. As most sample lines operate at or above ambient temperature, it is often necessary to insulate and even heat-trace sample lines, sample valves, filters, regulators, or any other components of the sample transport system where condensation or freezing may occur.

5. STEAM TRACING VERSUS ELECTRIC TRACING OF SAMPLE LINES

Steam tracing is the most popular and economical means of maintaining sample lines at elevated temperatures, particularly in locations where steam is a by-product of other process operations. A desired sample stream temperature is maintained by pairing the sample tube with a steam-transport tube and wrapping both lines together with insulation. The steam-trace tubing is equipped with valves, reducers, and steam traps and supplied with 15- to 230-psi steam, which correspond to temperatures of 250 to 400°F, respectively. As a general rule, larger-diameter steam-trace lines allow more-uniform temperature control and effective steam tracing over longer distances.

One common problem with steam-traced lines is "tailing," where the temperature of the sample line decreases with increasing distance from the steam source. This tailing is a result of heat losses both to the sample line and through the insulation material. The effects of tailing can be reduced by increasing the number of wraps of steam-trace tubing per unit length of sample line with increasing distance from the steam source.

A desired temperature setting is maintained by the use of a pressure reduction valve, where the greater the restriction the greater the pressure drop and the lower the resultant steam and sample temperatures. The desired temperature is maintained by the use of steam that is relatively free of scale and other contaminants, in combination with a steam trap whose purpose is to maintain steam flow and to remove condensation that forms as the steam temperature drops due to the previously discussed heat losses. In order of decreasing preference, the three types of steam traps commonly encountered are the inverted bucket, thermodynamic, and thermostatic types. The inverted-bucket steam trap is preferred because all of the condensate is collected within the bucket itself, not within the steam-trace tubing, as is the case with both the thermodynamic and thermostatic steam traps.

Electric tracing of sample lines is rapidly gaining popularity because it offers exact, uniform temperature control in a broad temperature range from 100 to 1000°F. Heating is accomplished by using either resistive or inductive heating elements. The available resistive heating lines use either series or parallel circuitry. The series-resistance electric tracing uses a high-resistance wire such as Nichrome for heating. The parallel-resistance heater elements use either resistance heating wire or an electrically conductive compounded material. The resistance wire heaters are constant-wattage elements with precalculated heat outputs. The conductive compound heaters vary in resistance in proportion to the generated temperature. As the temperature rises, the resistance of the compound increases until a maximum temperature is reached. The maximum temperature of these "self-limiting" heat-trace lines is determined by the particular formulation of the conductive material. Induction heating is used primarily for maintaining long sample lines at temperatures above freezing by using the sample line as a return for the heater circuit.

The most popular choice of electric tracing is the series-resistance type equipped with a variable transformer to adjust line voltage to supply the exact amount of current necessary to maintain the desired temperature. Also popular when temperature control is not as critical are constant-wattage heaters; the sample-line temperature is monitored, and the line voltage is switched on and off to maintain a desired temperature range.

You may be puzzled as to which type of tracing (steam or electric) to use, but the choice may have already been made for you. If the traced sample line is to be installed in a flammable or explosive atmosphere, classified as Class I Division I by the National Electric Code (NEC), then you must use steam tracing or another nonflammable heat-transfer medium (National Fire Protection Association, 1984). In areas classified as Class I Division II it is permitted to use either steam or mineral-insulated, constant power or self-limiting electrical elements for tracing.

B. Sample-Conditioning Systems

A properly designed sample-conditioning system is key to the success of a process-analyzer installation. Most analyzers will operate effectively if they are provided with a clean, dry, noncorrosive, homogeneous sample that represents the process stream being monitored; however, this is not typical of the sample provided to the sample-conditioning system of a process analyzer. Following the steps outlined in this chapter for sample takeoff and sample transport will result in the delivery of a representative sample, containing trace particulates, water, and other impurities, to the sample-conditioning system. The purpose of sample conditioning is to remove these last traces of undesirable materials to maintain levels that do not interfere with analyzer performance.

1. MATERIALS OF CONSTRUCTION

A sample-conditioning system must remove undesirable contaminants from the sample stream while remaining relatively inert to reactive materials present in both the process variables being monitored and the stream contaminants. The majority of sample-system components available today, including pressure regulators, valves, fittings, filters, rotameters, and gauges, are fabricated from 304 or 316 stainless steel or specialty alloys such as Hastelloy C to provide the desired levels of chemical resistance. Sealing materials are also metal, Teflon, or chemical elastomers, exhibiting similar resistance to chemical attack. Metals like copper, brass, and carbon steel are to be avoided in most applications. Most vendors of sample-system components can provide charts outlining the chemical-resistive properties of their products. These charts should be used to identify acceptable materials for construction of the sample system. Be certain to verify that the binding materials in glass and paper filters are also unreactive to your sample stream.

2. SAMPLE-SYSTEM COMPONENTS

Sample filters are used in series or parallel configurations in most sample systems. Filters in series typically consist of similar-sized filter bodies with coarse and fine filter elements. The coarse filter removes the bulk of particulates and water (coalescing filter), and the fine filter removes remaining traces of impurites. The coarse filter may require changing more frequently than the fine filter.

Many filter systems are designed with built-in redundancy to facilitate filter maintenance without interrupting analyzer operation. A bypass valve allows maintenance personnel to switch to a secondary set of filters while the primary set is cleaned or replaced. Another useful configuration is to insert the first in a series of filters in a sample loop near the analyzer. Most of the sample stream is bypassed from final filtering, and the sample stream is allowed to flow by to "cleanse" the outside of the first filter. Only the sample required for analysis is drawn through the first (and possibly a second) filter.

Sample regulators are used to control pressure in both sample loops and analyzer sample streams. Vaporizing regulators accomplish vaporization of a pressurized liquid stream to a completely gaseous stream at the sample takeoff point, offering flexibility in transporting a sample stream. Back-pressure regulators maintain the return sample pressure from an analyzer or sample loop at a level above its bubble point, eliminating undesirable flashing of volatile materials.

Sample rotameters are used to indicate the flow rates of sample and bypass streams. Special armored flowmeters are available for high-pressure and safety applications. Note that the rotameter float and barrel are calibrated for a gaseous or liquid stream of a particular density. A flowmeter with a scale in cubic centimeters of water per minute cannot accurately display the flow rate of most hydrocarbons. The following calculation can be used to determine the flowrate equivalent for any rotameter calibrated for a different liquid or gas stream:

$$\begin{matrix} \text{actual} \\ \text{flow rate} \\ \text{of} \\ \text{monitored} \\ \text{stream} \end{matrix} = \begin{pmatrix} \text{flow} \\ \text{indication} \\ \text{on} \\ \text{rotameter} \\ \text{scale} \end{pmatrix}$$

$$\times \left\{ \left[\frac{(\text{density of float}) - (\text{density of sample stream})}{(\text{density of float}) - (\text{density of material indicated on rotameter})} \right] \left(\frac{\text{density of material indicated on rotameter}}{\text{density of sample stream}} \right) \right\}^{1/2}.$$

Pressure gauges are available in both air- and liquid-filled varieties. The liquid-filled gauges are used to dampen pressure pulses encountered in some process streams. Snubbing devices are also available that can dampen pulses to standard gauges or can be used in combination with liquid-filled gauges for optimum pressure indication and increased gauge life.

3. ROUTINE MAINTENANCE CONSIDERATIONS

The maintenance advantages of selective sample-system redundancy have already been discussed as a means of increasing the on-line operating factor of an analyzer. Accessibility of sample-system and analyzer components that may require future maintenance should be considered in the initial system design. Provisions to introduce a calibration standard with known concentrations* of one or more of the measured variables will be used on a daily to weekly basis at the time of installation and startup and with decreasing frequency as a confidence level for the process-analyzer data is established. The calibration will be rechecked when the results from several process samples submitted to the laboratory for analysis do not agree with the data reported by the process analyzer.

Whenever possible, the process analyzer should be equipped with the necessary hardware to collect an analyzer-stream sample for laboratory analysis. Depending upon the physical properties of the sample, it may be necessary to collect the sample as a liquid or gas, often at conditions of elevated temperature and pressure. A Monel, stainless steel, or Teflon-lined sample bomb is frequently used in the collection of both gas and liquid samples for process-analyzer validation. The validation samples should be collected after sample conditioning and be as identical as possible to those analyzed by the process analyzer. This means of cross-validating process analyzers can be just as useful as the on-site calibration standard.

IV.   Process Analyzer Results versus Lab Results

When discrepancies arise between two analyzers that should be reporting the same results, the performance of the process analyzer is usually the first to be questioned. In some cases, recalibration of the analyzer will resolve the discrepancy. Occasionally, a process analyzer may uncover an unknown problem with a laboratory analyzer or its calibration standard.

What can you do if there appears to be no mechanical or calibration problems between the analyzers in question? Another possibility to consider, especially on nonroutine analyses or a new analyzer

---

* Since most facility laboratories are used to certify both purchased and manufactured products, every calibration standard, whether prepared internally or purchased from an outside vendor, should be checked in the laboratory.

TABLE II

CONVERSIONS BETWEEN LV%, WH%, AND MOL%

Component	LV%		Component specific gravity		Step	$100/\Sigma_1$	Wt%		Component molecular weight		Step	$100/\Sigma_2$	Mol%
(C$_2$H$_6$) Ethane	0.10	×	0.377	=	0.038	× 1.72 =	0.07	÷	30.07	=	0.0023	× 57.47 =	0.13
(C$_3$H$_8$) Propane	4.90	×	0.508	=	2.49	× 1.72 =	4.28	÷	44.11	=	0.10	× 57.47 =	5.75
(C$_4$H$_{10}$) n-Butane	95.00	×	0.584	=	55.48	× 1.72 =	95.43	÷	58.13	=	1.64	× 57.47 =	94.25
					$\Sigma_1 = 58.01$						$\Sigma_2 = 1.74$		

installation, is the units in which the results are reported. Do not accept percentage values without determining what kind of percentage. In the process industries, results are commonly reported in liquid volume (LV%), weight (Wt%), and mole (Mol%) percentages and unfortunately are often reported as "percentage" for convenience. The differences in results for a sample can be as large as 50% of the values reported! The procedure to convert from LV% to Wt% to Mol% is outlined in Table II for a three-component mixture of ethane, propane, and normal butane. This conversion technique is applicable in a similar fashion to any number of components. Comparing the results shown in the table we have the following summary.

Component	LV%	Wt%	Mol%
Ethane	0.10	0.07	0.13
Propane	4.90	4.28	5.75
n-Butane	95.00	95.43	94.25

V. Process Analyzer Shelter Specifications

The process analyzers of today play an increasingly important role in the areas of closed-loop unit control, analysis time and cost reductions, and compliance with newly enacted federal and state government regulations. Controlling the environment surrounding these analyzers is of critical importance, particularly in protecting modern analyzers and accessories that often contain delicate, solid-state microprocessor controls and a host of other electronic components.

A. Environmental Control

All types of analyzers and instrumentation are susceptible to deterioration if they are allowed to come in contact with their natural enemies, including rain, snow, ice, wind, sand, and dust. Over time, alternating periods of heat–cold and moisture–dryness cause expansion and contraction and lead to erosion and corrosion. Process-industry environments are particularly harsh, since rain and atmospheric moisture react with trace hydrocarbons, sulfur compounds, and nitrogen oxides to form acids, which accelerate corrosion.

B. Shelter Location

The selection of an analyzer site should reflect, in addition to the environment, the previously discussed considerations of area classification, shelter accessibility for placement and hookup, sample takeoff and return points, area availability of utilities including instrument air, water, steam, and electricity, and distance from the shelter to the point of analyzer output display.

The NEC criteria for classifying electrical requirements and areas of hazardous use has been adopted by OSHA as the minimum requirements for all electrical installations, and compliance is mandatory.

C. Shelter Selection

If you must install process analyzers in an NEC-classified hazardous area and a shelter is required, you have two possible approaches to consider. The first is to reduce or eliminate the hazardous-area designation *inside* the shelter. This can be accomplished by installing an elaborate shelter equipped with double, air-lock entry doors and a failsafe system to automatically disconnect electrical power in the event of a loss of shelter pressurization. This method of installation is very costly and should be considered only in extreme cases where there is no suitable process analyzer available for hazardous-area installation or when one must install an analyzer with a foreign safety certification or a domestic analyzer in a foreign country. No international code exists for analyzer area certification. A useful guide covering international practices for electronic monitoring equipment is listed in the references and may prove valuable on a foreign assignment (Comins, 1980).

The preferred approach to meeting NEC requirements is to obtain analyzers and associated devices that meet the particular hazardous-area requirements and to install them inside of a good-quality, general-purpose shelter. Desirable shelter features include adequate working area, centrally located floor drain, ventilation with manual damper closure, circuit breakers in lieu of disconnect switches, and a coating of an epoxy-based enamel paint such as DuPont Imoron on exposed metal surfaces. Shelters should be anchored to a solid concrete foundation, and the shelter pad should be poured as soon as possible to "reserve" your optimal shelter location from alternative use.

References

Comins, C. C. (1980). *InTech,* October.
Houser, E. A. (1972). "Principles of Sample Handling and Sampling System Design for Process Analysis." Instrument Society of America. Pittsburgh.
National Fire Protection Association (1984). National Electrical Code NFPA, No. 70. Boston.

16

Engineering of Total Analyzer Systems

GEORGE F. ERK

Sun Refining and Marketing
Philadelphia, Pennsylvania

I. Introduction

On-line process analyzers have become an important tool in the process industry because they can improve production and plant operation and reduce or even eliminate out-of-specification products. When on-line analyzers first came into use in the 1950s, each continuous process-stream analyzer was handled on an individual basis. An engineer strictly evaluated the suitability of each analyzer application. Armed with specification sheets, certified prints, and instruction manuals, all individual components were engineered and designed into a total system. Old, abandoned pump houses, storage rooms, or perhaps even a new modular sheet-metal building were used. The engineer assembled all the components within that enclosure, and the system was constructed at the final job site. Finally, the completed assembly was calibrated and checked; if everything worked, then it was put on-line to produce results.

This rather cumbersome and random approach worked quite satisfactorily as long as single applications were involved. Analyzers were primarily installed into existing process plants, some on a trial-and-error basis. Special problems that arose frequently were all handled by people who had some knowledge and gained experience as they worked on the problem solving. However, for new plant construction and for the simultaneous installation of several analyzers, a better approach was developed in the mid-1960s. At that time many installations saw as many as 50 or 100 continuous process-stream analyzers being engineered on a single project. Modular and preassembled analyzer systems were installed with modular sample collection and treatment devices and standardized calibration and sample-disposal methods and were preassembled into a dedicated shelter at a manufacturer's location. Advantages of modular, preassembled systems include the following.

(i)   Systems engineered, designed, and manufactured by a specialist supplier are often superior to those built on the job site.

(ii)   Construction in the factory is independent of weather and labor conditions at the job site. This may reduce installation cost and, perhaps more importantly, may guarantee a reliable time schedule.

(iii)   The entire system can be fully tested under simulated operating conditions, and any design, equipment, and construction faults can be corrected before shipment is accepted by the customer.

All of this is accepted practice today. However, many management and engineering details have to be planned throughout the entire project. A typical sequence of events occurring on such a project is presented below.

II.   Analyzer Systems Planning

A.   Development of Analysis Requirements

The first task that faces the analyzer engineer is review of the functional process requirements; from this the type of analysis that is best applied is determined. A thorough review of the process data, gathering representative information and determining worst-case conditions on either side of the normal measuring range, is necessary before the analysis requirement can be defined. A process specification that spells out the mole percentage of each component to three or four decimal places may be totally useless if the deviation ranges and maximum and minimum extremes possible during any type of operation, including upset, start-up, shutdown, and turnaround times (analyzer-out-of-service conditions), are not listed and known to the designers. Temperatures and pressures identifying the state of all components must be obtained, again under all possible conditions; it is very important to know if a component is in the liquid phase, in the vapor phase, or at the bubble point.

B. Evaluation of Methods

Once process conditions are known, an analysis method must be selected to handle the application. The analysis method must fit all process conditions and not vice versa. Factors that can influence the selection of an analysis method include cost, response time, maintainability, reliability, ease of operation, and ease of calibration.

C. Justification of Analyzer Installation Cost

Analyzer installations not required for safety and environmental protection generally have to be justified on an economic basis. A simple computer program can be established to calculate investment parameters (the output of such a program is shown in Fig. 1) based on the amount of fuel saved, the increase in product yield, the higher efficiency of operation, or the decrease in or elimination of out-of-specification material. Even analyzers for detection of hazardous conditions might find economic justification for their installation, since they may prevent costly and time-consuming emergency shutdowns of process plants.

As additional examples of economic justification, Figs. 2–5 represent studies performed on an oil-refinery computer model developed for the U.S. Department of Energy by Turner Mason and Associates. These figures show the payouts on viscosity, boiling-point, and pour-point analyzers.

Computer input data

Heater size: 30 MMBTH
Heater efficiency: 88%
Fuel heat value: 6x10⁶ Btu/FUEB
Yearly operating time: 8,400 h
Increased heater efficiency: 1%
Analyzer installed cost: $11000.
Yearly maintenance cost: $2000.
Investment tax credit: 10%
Analyzer life: 10 years

Computer output data

Total investment = $11000.
Taxes = 50.80% paid 4 times/year
Project Life = 10.00 years
Lead time = 0. years
Undepreciated book value at end of project = $0.
Interest rate of return = 22.75%
Total return on investment = 126.29%
Average return on investment = 12.63%/year

Discount rate	Present worth
0.00%	$13892.
10.00%	$ 5561.
15.00%	$ 2918.

Payout (years)
3.64
4.68
5.66

End year	Invest-ment	Book value	Deduct. expense	Depreci-ation	Taxes	Investment tax credit	Cash flow
0	11000	11000	0	0	0	0	0
1	11000	8800	3700	2200	762	1100	4038
2	11000	7040	3700	1760	986	0	2714
3	11000	5632	3700	1408	1164	0	2536
4	11000	4506	3700	1126	1307	0	2393
5	11000	3604	3700	901	1422	0	2278
6	11000	2884	3700	721	1513	0	2187
7	11000	2163	3700	721	1513	0	2187
8	11000	1442	3700	721	1513	0	2187
9	11000	721	3700	721	1513	0	2187
10	11000	0	3700	721	1513	0	2187
Total	11000	0	37000	11000	13208	1100	24892

Fig. 1. A computer program for investment parameters.

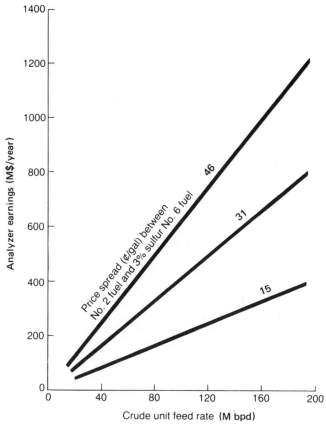

Fig. 2. Estimated profitability of a viscosity analyzer on a crude unit. The analyzer earnings were computed after maintenance costs were deducted. Feed rate is given in millions of barrels per day. (By permission of GCA/Precision Scientific Group.)

D. *Establishing Typical Scheduling Sequence (Planning Diagram)*

While items discussed up to this point need to be planned, one must also thoroughly evaluate the sequence of events in the engineering and management of the project as well. The planning diagram shown in Fig. 6 shows all the major activities during the entire project. Starting with the previously mentioned definition of requirements, through evaluation, economic justification, specification development, bidding, manufacturing, site preparation, shipping, in-

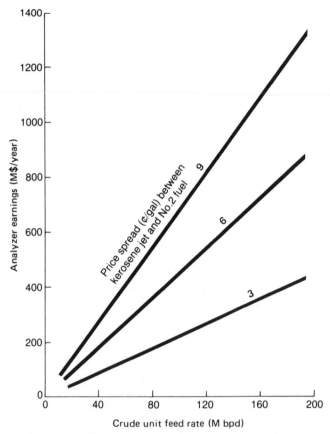

Fig. 3. Estimated profitability of a single boiling-point analyzer on a crude unit. The analyzer earnings were computed after maintenance costs were deducted. (By permission of GCA/Precision Scientific Group.)

stallation, checkout, commissioning, and post–start-up support, all items are important to the satisfactory completion of the entire project. Depending upon the application, certain items can be accomplished at different times. Likewise, the sequence of events does not imply a degree of importance for any one factor.

E. Preparing Analyzer Systems Specifications

When analyzer requirements and the method of analysis have been decided, the sample point must be located. Process conditions

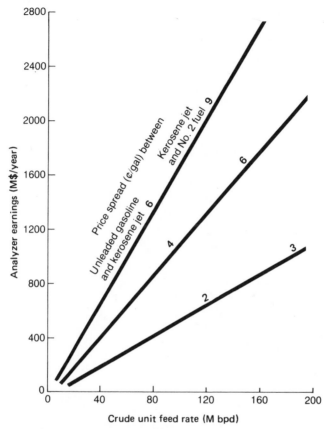

Fig. 4. Estimated profitability of four boiling-point analyzers on a crude unit. The analyzer earnings were computed after maintenance costs were deducted. (By permission of GCA/Precision Scientific Group.)

at the sample point, such as temperature, pressure, viscosity, pour point, phase, impurities, and component ranges must be established. The sample point should be located so as to have good access to valves, regulators, coolers, and other process plumbing and also to minimize time response of the system. Longer lines from the sample point to the analyzer increase the time lag. Engineers must decide whether to take samples from horizontal or vertical lines and whether to take samples from the top or side of a line.

As a typical example of what faces the engineer, consider a sample requirement to measure the overhead material of a distillation column. Ideally, it would be desirable to take the sample directly

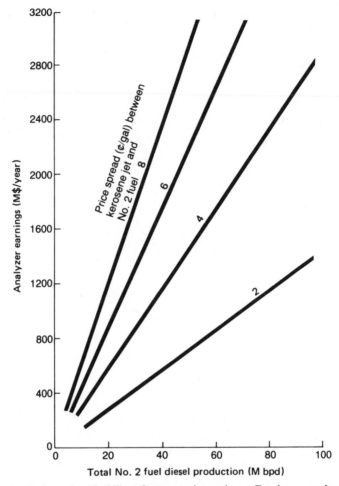

Fig. 5. Estimated profitability of a pour-point analyzer. Earnings were based on benefits during four winter months only and were computed after maintenance costs were deducted. (By permission of GCA/Precision Scientific Group.)

from the top tray of the tower; however, industrial distillation columns with dozens of trays could be several hundred feet high and very difficult to access. A location close to the overhead condensers could provide a similar sample; but even overhead condensers are at least three floors up in the steel structure holding them and the accumulators (overhead product-receiver vessels), and since most maintenance organizations prefer to have analyzer systems located at ground level, it it still a long distance from the overhead con-

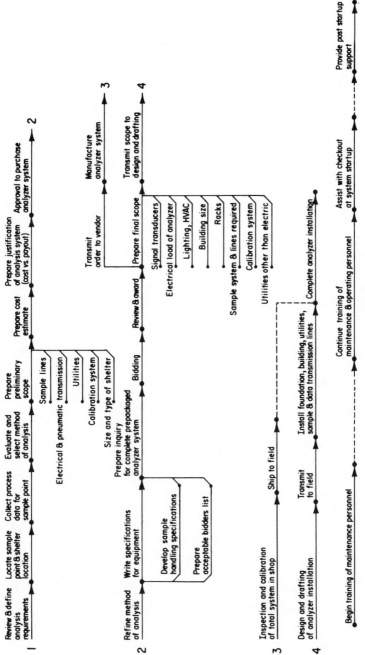

Fig. 6. The planning diagram.

denser inlet to the analyzer building, requiring a lengthy section of tubing. When the overhead is a final specification product in its liquid form, measuring the liquid composition would eliminate measuring any noncondensables; since they are vented off the accumulator, they would not have to be analyzed. If the liquid is pumped to storage, there is the added possibility of using the pump head and a "fast loop" for sampling and disposal of spent sample.

All the data (process conditions, sample location, and analysis requirements) should be entered on a data sheet that later can be used for bid requests and purchase orders. A sample of an analyzer data sheet is shown as Fig. 7. After completion, this data sheet would contain information on (i) analytical requirements, (ii) readout requirements, (iii) sample stream composition, (iv) sample stream condition (phase, temperature, pressure), and (v) required electrical hazard classification. Factors that are frequently overlooked in the preparation of such a data sheet include the following:

(i)   temperature, pressure, and stream composition maxima or minima that may be present during process upsets, start-up, shutdown and the like;
(ii)   trace contamination of compounds that are corrosive to a sample-handling system or may interfere with the analyzer measurement;
(iii)   documentation required with analyzer; and
(iv)   special fittings or type of material preferred because of plant standards or existing spare parts stocks.

1.   SAMPLE HANDLING

The purpose of a sample-handling system is to deliver to the analyzer a clean sample that is fully representative of the process stream. A sampling system includes all piping and equipment necessary to transport the sample from the sample point to the analyzer and then to a return or disposal system. (Particular care must be exercised to avoid air- or water-pollution problems). Particularly difficult design problems arise in processes that involve mixed-phase systems and heat tracings. In applications where heat tracing is critical, electric tracing offers closer temperature control than does a steam system.

In the case of distillation towers with relatively simple control systems, one analyzer should be dedicated to a single stream. If more than one stream per analyzer is sampled and they are alternately measured on a fixed cyclic basis, it has to be remembered that

	TAG NUMBER		
	SERVICE		
ANALYZER	TYPE ANALYZER		
	ANALYZER LOCATION		
	MOUNTING		
	ACCURACY		
	SENSITIVITY		
	RESPONSE TIME		
	ELECTRICAL CLASSIFICATION		
	ANALYZER MODEL NO.		
CONT.	CONTROL UNIT LOCATION		
	MOUNTING		
	ELECTRICAL CLASSIFICATION		
	CONTROL UNIT MODEL NO.		
MFR.	MANUFACTURER		
	PURCHASE ORDER NO.		
SAMPLE SYSTEM	SAMPLE TAP LOCATION		
	SAMPLE SYSTEM LOCATION		
	SAMPLE LINE LENGTH		
	SUPPLY PRESS. MIN. | MAX.		
	RETURN PRESS. MIN. | MAX.		
	VENT PRESS. - LOCATION		
	DRAIN PRESS. - LOCATION		
	ELECTRICAL CLASSIFICATION		
	CALIBRATION STANDARD		
	VENDOR		
	PURCHASE ORDER NO.		
READOUT	INDICATOR MODEL NO.		
	RECORDER MODEL NO.		
	MANUFACTURER		
	PURCHASE ORDER NO.		
SHELTER	SIZE AND TYPE		
	ELECTRICAL VOLT - FREQ.		
	INSTRUMENT AIR PRESSURE		
	WATER PRESS. | TEMP.		
	STEAM PRESS. | TEMP.		
	CARRIER GAS TYPE | PRESS.		
	AMB. TEMP. RANGE		
	TYPE WINTERIZING		
FLUID DATA	NO. STREAMS MONITORED		
	COMPONENTS MEASURED		
	SPANS		
	FLUID | PHASE		
	TEMP. MAX. | MIN.		
	SP. GR. 60°F | N.O.T.		
	VISC. 60°F | N.O.T.		
	COMPONENT CONCENTRATION MOL. % WT. % MOL. % WT. %		
	1 MAX.-MIN.		
	2 MAX.-MIN.		
	3 MAX.-MIN.		
	4 MAX.-MIN.		
	5 MAX.-MIN.		
	6 MAX.-MIN.		

Fig. 7. A typical analyzer specification sheet.

the additional memory and control logic will make the system somewhat more complicated and less reliable. Recent microprocessor-
based–logic control systems have improved such control sequenc-
ing tremendously over mechanical sequencers and counters. If
switching is required, care must be taken to prevent cross-contami-

nation. Devices such as air-operated double-block-and-bleed valves are applicable here.

If more than one analyzer is to be provided for one process unit, an entire unit operation, or even a single point on the process, then all these analyzers should be arranged into convenient groups within a single shelter. Grouping of analyzers permits easier maintenance and calibration of the analyzers. When several analyzers are housed in a single shelter, the installed-unit cost goes down, because the analyzers share common facilities such as heating, lighting, air conditioning, and other utilities.

Another major design consideration is to build the sample-handling system around individual, single-purpose, discrete components that can be easily interchanged or replaced. This simplifies repair and maintenance.

2. ANALYZER REQUISITIONS

The central document for the entire project is the analyzer requisition form. It consists of functional specifications and describes what is required of the analyzer systems. It also describes the interfacing of the vendor with the engineer and/or the ultimate user. It should contain a timetable for progress reports, payments scheduling, and progress payments, if any, and it should state all the commercial terms and conditions that your company has established for doing business with suppliers. The requisition form should also indicate to the vendor all the available utilities at the site and request information about what requirements the vendor has. Also normally included in the requisition package would be a drawing showing the plot plan of the shelter in the process units and the location of the sample connection.

3. PIPING, WIRING, AND UTILITY REQUIREMENTS

a. Piping considerations. In the chemical and petroleum industries many process streams consist of highly flammable materials. If excessive quantities of flammable sample are brought to the analyzer shelter to minimize lag time, the by-pass plumbing (fast loop) manifold should be located outside of the shelter. This will prevent the sample from entering the shelter if a leak occurs in the high-pressure, high–flow-rate sample line. Flammable calibration fluids should be stored outside the shelter and protected from the sun and hostile environments (rain and snow).

High-pressure and high-temperature lines should also be routed outside the shelter for safety reasons, and piping connections inside the building should be minimized. Tubing should be used wherever possible.

Exceptions to these rules are low–flow-rate gas stream-switching systems which usually are brought into the shelter to prevent condensation. Hot lines, which must be located in the shelter, should be insulated and heat-traced as required. Hot lines that cannot be insulated should have guards around them to prevent injury to operating and maintenance personnel.

Piping should be arranged to provide ease of calibration and maintenance. Sloping lines can be used to prevent condensate trapping, since condensate formation cannot be tolerated. In some cases, heat tracing may be required. Effluent lines from the analyzer condensation traps should be sloped to facilitate condensation removal.

b. Wiring considerations. All field wiring should be connected to the appropriate terminal strips in terminal junction boxes that are mounted outside on the shelter. Wiring from these to all required locations in the shelter is then installed at the factory, and it should be designed so that all interfaces to the process-plant wiring are accomplished through these externally mounted terminal junction boxes. Particularly in the case of new plant construction, the analyzer shelter should be designed so that field craft personnel do not require access to the interior to make electrical or piping connections. Analyzer shelters should be shipped to the field locked, and they should remain unopened until start-up begins.

If various classes of wiring are entering the shelter, a separate terminal box should be provided for each; dc wiring should be run in separate conduits from ac wiring, and spacing between ac and dc wiring should be at least 12 in. Analyzer output and control signals must be considered in the electrical design. For example, if the analyzer output is a dc millivolt signal and the control logic is ac, then the wiring cannot be incorporated in the same conduit. Two conduit runs would be needed between the analyzer and its control section. Although many analyzers have ac control signals, modifying the unit to accept dc control logic could be investigated. If the distance between the control section and the analyzer is great, then conduit cost could be a deciding factor in analyzer selection.

Qualifying an analyzer for Division I requirements is both an economic and engineering decision. Many analyzer enclosures are qualified for Division 2 areas; but when hydrocarbon vapors, or even hydrogen, are present and the area is rerated Division I, problems arise. The enclosure can be air-purged, in combination with several interlocks, but these methods are complex, expensive, and difficult to maintain.

Analyzer packaging can be a factor in wiring design. Certain analyzers and controls are packaged in a single explosion-proof box; others are furnished in a combination of general-purpose and explosion-proof enclosures. Interconnection costs are much less with a single-enclosure system.

In general, all electrical systems must be in accordance with applicable electrical codes in the geographical area of the plant; this includes the NEC, state, county, city, or any other local codes that may apply.

c. Utility requirements. The utility requirements for the entire system must be carefully documented and consolidated for the job-site construction contractor. Information is available from the analyzer instruction manuals, various sample-system specifications, and other technical data related to equipment that is required inside the shelters (i.e., electrical or steam heaters, air conditioners, and lighting equipment). Disposal requirements (drainage and exhaust) also need to be documented. Since these utilites must be supplied at the jobsite, it is important to establish these requirements early in the construction project.

4. READOUT DEVICES

Conventional readout devices can be utilized to present most of the analyzer data. For example, temperature readout (boiling-point unit) may go directly to a multipoint temperature recorder and be printed out together with other temperature measurements on the process. Indication only may suffice in some areas; alarms incorporated into an indicator may just alert the operator to a condition that requires attention. On the other extreme, sophisticated computer print outs that provide detailed information, such as date and time of present reporting, average reading of recent past (hour, shift, day, week, month), and deviation from normal, may be specified per request from operators, yield departments, technical services personnel, etc.

If analyzer manufacturers do not offer a standard readout device within their package, most conventional readout devices could be specified as long as a standard signal output (such as 4–20 mA) is provided by the analyzer manufacturer.

When connecting a transmitting signal from the analyzer directly to a computer, a number of engineering factors have to be considered. If the analyzer is not equipped with a computer interface, then a register and sample-and-hold memory device may be needed to properly identify the analyzer and stream. The signal interface to the computer does require a more carefully designed wiring system. It may require shielding to prevent plant electrical noise from interfering with the data, or it may require special signal conditioning to convert the data into a computer-usable form.

Today, under many conditions, dedicated microprocessors are used for individual analyzers or groups of analyzers; these microprocessor systems could be independent subsystems able to interface with other hierarchical computers. These microprocessors are programmed to provide functional control commands, such as stream switching, and they have the interface logic for communicating with other computers. To minimize complexity with any computer interface, all analyzers selected should provide some standard output signal, such as a 4–20 mA or 0–10 Vdc.

If a microprocessor or microcomputer is selected and no permanent analog recording is provided, it may still be necessary to provide analog recording (trend recording) for calibration and maintenance. Conventional analog recorders in the maintenance shop or portable devices may be utilized for this purpose.

5. Environmental Conditions

When environmental conditions are not taken into consideration during the design of the system, inefficient operation or even erroneous measurements may result. For example, an analyzer exposed to direct sunlight on a hot day may not operate within the published accuracy range. It is therefore important to make sure that the environment is kept within the limits specified by the manufacturers of the analyzers. Also, a minimum temperature may need to be specified to prevent freezing of water, even if the analyzer may be operational below the freezing point of water.

If natural-draft roof ventilators are used, sufficient air openings are required on the bottom of the shelter to allow an unobstructed flow of air. Some designs force air in from the top of the shelter. The air flow should be based on whether the gaseous material escaping from the analyzer system is mostly lighter or heavier than air.

6. TRAINING

As Fig. 6 shows, training begins long before the analyzer system is complete and ready to start. Maintenance people should get involved during inspection and calibration of the total system in the fabrication shop. This would allow them to suggest changes that would improve the installation. Operating personnel should likewise get involved during field installation for the same reason. While it would be ideal for both groups to make all their suggestions during the initial reviewing period, it is more realistic to provide for at least some last minute input; changes may be totally prohibited after the installation is complete.

7. CALIBRATION

A properly designed, effective calibration system could contribute an additional 20–50% to the analyzer cost, and it is an important factor to consider in analyzer selection. Calibration is usually accomplished by adjusting zero and span settings to agree with values obtained from manual introduction of calibration gases or liquids from pressurized cylinders.

Another calibration method, more costly and time-consuming, is to take several "grab" samples and perform laboratory analyses; results of the analyses are then compared with analyzer readings. However, to have statistical confidence in the data, several grab samples should be evaluated, and the laboratory's standard deviation must be known.

A more sophisticated calibration method is performed in conjunction with a computer. With this method, certified calibration liquid or gas samples are introduced into the analyzers on a predetermined automatic basis; the computer then evaluates the analyzer output data against known values of the calibrated sample. After calibration, the computer uses the information to calculate output drift and correct future analyzer outputs. Also, the information can be used to inform maintenance personnel that an analyzer should be taken out of service for repair and cleaning or field adjustments. Some analyzers are more adaptable to an automatic calibration technique than others; if this procedure is contemplated, then the engineer must consider such adaptability in his analyzer selection. Also, calibration gases stored in cylinders can change composition (stratification), and this should be evaluated when considering automatic techniques.

8. DOCUMENTATION

Analyzer specifications, outline drawings, installation drawings, and manufacturer literature (such as operating and maintenance manuals) all must be kept as permanent records. In most cases several sets of these documents are required and should be requested in the bid invitation as well as specified in all details in the procurement documents.

III. Analyzer Systems Manufacturing

Reviewing acceptable fabrication shops, getting proposals, and placing an order for a prepackaged analyzer system is not much different from any other materials-management procedure. It is more complicated, however, than placing an order for a field transmitter or a control valve because of the diversified aspects involved. But the number of reputable "system houses" in this field has grown over the past few years, and a good choice of companies is available.

The analyzer system house is usually responsible for installing analyzers from various manufacturers into the analyzer shelter. They install all the wiring, plumbing, and utilites inside the shelter according to specifications, build and install the calibration system, and conduct various calibration and acceptance tests on the equipment. Most, if not all, of this work is conducted in a fabrication shop away from the plant site. Figures 8–17 illustrate typical analyzer systems during manufacture and after field installation.

During the manufacturing stage, it is important to periodically receive status reports, to make sure the time schedule is being followed. If personal inspection trips are too expensive, then instant-camera photographs of the work in progress may suffice. While the analyzer system is being fabricated, field construction work should remain on schedule so that no delays occur at the jobsite.

When the analyzer system is completed, an inspection and calibration is conducted at the fabrication shop. Calibration should be done using a certified sample that is representative of the process stream on which the system is to be used. This test should be run at least once; a burn-in period may also be specified in which the analyzer should operate for a specified length of time (24 h, 100 h, etc.) without any failures. If a failure occurs, then the test starts over from time zero.

Fig. 8. Typical frame for analyzer building.

Fig. 9. Analyzer house.

Fig. 10. Typical piping and wiring.

Fig. 11.   Process tubing.

Fig. 12.   Typical external piping.

Fig. 13. External tubing and piping.

Fig. 14. Completed analyzer building.

Fig. 15. Analyzer shelter.

When the analyzer system is officially accepted, the manufacturer of the analyzer system ships the shelter to the jobsite, usually on an air-cushioned flat-bed trailer.

IV. Analyzer Systems Installation and Calibration

On arrival at the jobsite, plant technicians install the shelter and tie it into the field wiring, piping, and utilites. If connections have been marked properly, then installation can be done in a minimum amount of time.

A final checkout in the field is usually done with another calibration sample. This checkout ensures that the system is still operational and that no damage occurred during shipment or installation. If training of operators and maintenance personnel was done properly and in advance of start-up, then operational support will be available from plant personnel immediately after installation.

Fig. 16. Analyzer building in operation.

Fig. 17. Plant analyzer building.

V. Analyzer Systems Maintenance and Training

There are several approaches to systems maintenance. The choice is between providing in-house personnel or contracting for this service; a combination of in-house and contract service could also be considered. Specialized contractors are available, and they provide the expertise "ready made," and, consequently, the end user need not be concerned with training. If management decides to do all maintenance with permanent employees within their own organization, then training becomes an important part of preparing the selected persons for the job at hand. A careful selection is first made among the generalist-type instrument people to identify technicians who would qualify from their past experience for this more elaborate task. Also, it has proven to be beneficial to put these people into a separate organization with a higher pay level. Requirements here are higher than for the instrument generalist, thus justifying higher status if qualified.

Training can also be accomplished in several ways. As shown in Fig. 6 and explained above (Section II.E.6), "on-the-project" training provides a valuable way to learn and understand a particular analyzer application. However, more-general basic training is also necessary and can be done by either supplier companies in the factory or by their qualified instructors at the jobsite. Likewise, trade schools and professional societies provide classes that are more "generic" in their coverage.

VI. Analyzer System Postaudit

When justifying an analyzer system and submitting it for an appropriation, all conditions are stated on a somewhat theoretical basis. Actual performance in the field is likely to be different from all the parameters used before the installation. Consequently, some time after the system has run under normal plant conditions, perhaps as much as a couple of years, a thorough review of the system should be done to verify the payout. Some installations may not achieve the original targets, others may have surpassed them. In any event, such a review of the project is not only desirable from a management viewpoint, but it will also provide for all people involved a better insight into what went right or wrong during the entire process. It will help in doing a better job in the future, even if the present system turned out to be a good one.

Bibliography

Bailey, S. J. (1983). *Control Eng.* **30,** 57–61.
Erk, G. F. (1977). *Instrum. Tech.* **24,** 39.
Jutila, J. M. (1980). *InTech* **27,** 16.
Krigman, A. (1983). *InTech* **30,** 9.
Puzniak, T. J. (1980). *Anal. Instrum.* **18,** 89–94.

17

Process, On-Stream, and Chromatographic
Measurements in Brewing

KARL J. SIEBERT

Research & Development
The Stroh Brewery Company
Detroit, Michigan

I. Introduction

Brewing is not a formulation process where components are simply mixed together to arrive at the finished product. Rather, it is a process during which many biological and chemical transformations occur. During the major transformation, the starch of the grains is enzymatically converted into simple sugars; these, in turn, are transformed by yeast into alcohol and carbon dioxide.

Because brewing developed over many years as an empirical discipline, scientific explanations for many phenomena did not come until more recently. This scientific knowledge has helped us to gain control over certain aspects of the process. Progress has continued and has changed brewing from mostly art to mainly technology, but it has not yet become a science.

Brewing began over 5000 years ago as a batch process. Although many attempts have been made to change beer production into a continuous process or a process with some continuous stages, only a very small percentage of the beer made in the world today is produced with either continuous wort production or continuous fermentation. In part this situation stems from incomplete knowledge of all of the substances important to key properties of beer. Where knowledge is adequate, the complexity of material and process-step interactions has made it difficult to duplicate existing products by the use of continuous process stages.

The properties with which the brewer is most concerned are, of course, those that can be perceived by the consumer. In the main these are flavor, foam, color, and haze. A number of additional substances and properties are of interest either as predictors of these four properties or for control of the process in terms of uniformity and costs.

By far the most important and complicated of these properties is beer flavor. Unlike many process situations where the object is maximization of yield of a single defined compound, the object of brewing is the maximization of "good flavor." Beer flavor is particularly complex, depending on two moderately strong influences, ethanol and hop bittering compounds, and a great many characteristics that are more subtle. Of the over 850 compounds known to be present in beer (Meilgaard, 1982), approximately 40–50 are thought to be normally present in flavor-active amounts. Together with ethanol and the bitter substances, these probably account for only about 60% of normal beer flavor. Understanding of the situation is not aided by the fact that taste panels are subjective and nonreproducible "instruments." Recording the responses of individual tasters to various flavor substances in beer has shown that detection thresholds for a single compound may vary from person to person by a factor that is often 2–10 and occasionally up to 100 (Meilgaard *et al.*, 1982). It is normal for any one person to be very sensitive to some substances, moderately sensitive to most, and insensitive to others. The human olfactory organs respond very differently from most instruments. The average thresholds (of a group of tasters) for known beer compounds range from 1.4% (wt/vol) for ethanol to 10 ng/liter for dimethyl trisulfide, or over nine orders of magnitude. This means that even with the high degree of separation achievable with chromatographic methods, many very large peaks, most of which are caused by compounds that do not produce perceptible flavors, will tend to hide the smallest peaks, some of which are due to flavor-significant compounds. While chromatographic methods generally offer the best approach to flavor research and can in many cases be adapted for on-line analysis, the problems with beer flavor analysis described above have for the most part prevented the on-line application of such methods in brewing. Chromatographic methods are frequently used in brewing laboratories for control of certain flavor properties, and it seems likely that some could be moved into the plant at some future time.

This chapter will describe the raw materials of brewing and the three main steps in the brewing process. Process control parameters and on-stream analyses will then be discussed. Finally, chromato-graphic methods used to monitor the process will be described.

II. Raw Materials of Brewing

Beer can be made with only barley malt, water, hops, and yeast; but most beers in countries other than Germany employ an adjunct, which provides an additional source of starch or sugar besides malt. In this chapter, beer is taken to include both ales and lagers. For greater detail on the brewing materials and process, "The Practical Brewer" (Master Brewers Association of the Americas, 1977) is recommended.

A. Water

Sterility and flavor purity are mandatory, and blending water must be oxygen free. Breweries have some of the largest water-treatment plants of any industry. It is common to pass the water through activated carbon and/or to treat it with ozone. Both methods re-move or alter organic matter that might otherwise lead to off flavors. Ozone is also a microbiological sterilant. Activated carbon treat-ment is necessary when the water has been previously chlorinated, since carbon also serves to remove chlorine. Lime softening or ion exchange is required for some waters. Deaeration is performed by vacuum treatment and/or by purging with air-free CO_2.

B. Malt

Malt, which has been called the "soul of beer," is produced from barley in the three-step process depicted in Fig. 1. Barley seeds are soaked or steeped in water, which is drained and replaced every 6–8 hr, with a total steeping time of approximately 1 day. This raises the moisture content of the seeds from near 12 to about 45%. Air is blown through the steep water to provide oxygen for respiration. Beginning with steeping and continuing through germination, the barley kernel starts to generate energy through respiration. At first this is mainly through lipid metabolism, but there is a gradual shift to carbohydrates as the energy source. The seed begins to break down storage protein into amino acids and uses these to synthesize en-

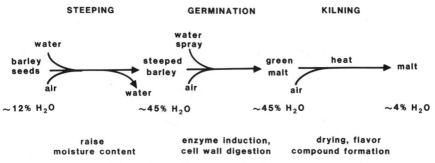

Fig. 1.   Flow diagram of the malting process.

zymes that were not previously present and will be needed for growth. The enzymes induced in this manner include α- and β-amylases, β-glucanases, and proteases, all of which are important for brewing.

After steeping, the grain is spread on a flat, slotted surface where it is kept moist with a water spray. Air is blown through the grain bed, and the grain is turned periodically. During this time the barley begins growing and starts developing rootlets and a sprout. Germination greatly alters the internal structure of the barley kernel; the enzymes dissolve cell walls in the grain endosperm and attack the starch granules in this tissue. This results in a physical change that renders the entire grain much softer and makes the starch more available than it was in the starting barley seed. At the end of germination, the moisture content of the green malt is essentially the same as it was at the beginning.

Heat and large quantities of air are passed through the green malt in a kiln in order to dry it. A temperature program is used that is designed to minimize the inactivation of amylase activity. This is done by employing moderate temperatures until the moisture content of the malt has been reduced. In the final stages of kilning, high temperatures (near 85°C) are reached. The flavor of malt is mainly developed during kilning. Reactions between reducing sugars and amino acids (Maillard reaction) as well as simple caramelization result in color formation and a spectrum of roasted-flavor compounds with the distinctive malt smell. Different kilning regimes lead to malts ranging from pale to dark, with flavors to match. The finished malt kernel is a stable, pleasant-flavored bag of available starch together with the enzymes needed to degrade it.

C. Hops

Hops grow on a vine like grapes, and their only commercial use is for beer production. Hops contain two main properties of interest to the brewer. The sharp, clean bitterness of beer is derived from the α-acids of hops. These compounds are produced in amounts ranging from 3 to 16% of the hop dry weight, depending on variety and growing conditions. The α-acids are not themselves bitter, but they are converted into the bitter iso-α-acids during the brewing process.

The essential oils of hops give rise to the hoppy aroma of beer. Few of the hop oil compounds, which are mainly of terpenoid origin, can be found in beer. It appears that some oxidation of these compounds into oxygenated forms is required to render them soluble enough in water to survive the brewing process.

Hops are simply dried and baled after harvest. The dried hops may be used as leaf hops, ground and pressed into pellets, or extracted with solvent or liquid CO_2. Hop products are stored cold until they are used.

D. Adjuncts

Most malts have more than enough enzymes to break down the starch contained in the malt kernel. As a result, an additional source of starch, called an *adjunct,* can be used in place of part of the malt. This produces a somewhat less satiating beer of lighter color and greater haze stability. The most commonly used adjunct is corn grits, followed by broken rice. Much smaller amounts of unmalted barley, wheat, or sorghum are used in some places and for some products. Less commonly, purified starches derived from cereal grains are employed.

Syrups, in which the starches have been broken down into simple sugars by acid hydrolysis, enzymatic hydrolysis, or combined acid and enzyme hydrolysis, can be produced from barley or, more often, corn. These do not need the action of malt enzymes and can be added later in the process than starchy adjuncts.

E. Yeast

Only a small percentage of the many known strains of *Saccharomyces carlsbergensis* (lager yeast) or *S. cerevisiae* (ale yeast) are suitable for brewing. Yeast is not a raw material in the sense of the other items listed above; it is a self-replenishing resource. The brewer harvests three to four times as much yeast from a fermenter as he puts into it. Brewers take great care to maintain their yeast free from bacteria and wild (nonbrewing) yeasts and to guard against mutation. Often this is done by periodically propagating enough yeast from a small, pure culture to use in a plant-scale fermentation.

III. The Brewing Process

The process of making beer is conceptually divided into three main areas, which are depicted in Fig. 2. In wort production, water, malt, adjunct, and hops are combined to produce a suitable nutrient medium for yeast. Flavors are derived from the malt and hops during this stage. Yeast is "pitched," or added to the wort, and causes the formation of ethanol, carbon dioxide, and many other aroma-volatile compounds. Yeast is removed at the end of fermentation, and the beer is lagered, or held cold, while colloidal stability is achieved and the flavor matures. Most beers are then filtered before packaging and sale. Cask-conditioned ales (usually only in Britain) are not filtered but are fined (by the addition of a coagulating agent), and time for settling is allowed before dispensing.

A. Wort Production

Wort production is the most intensive, hottest, and most rapidly conducted (approximately 10 h) stage of brewing. While brewhouse equipment varies in some details, the general concepts employed are

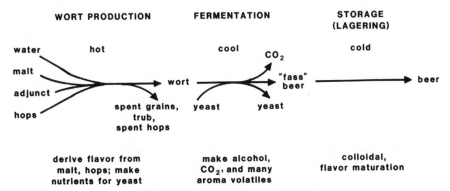

Fig. 2. Flow diagram of the brewing process.

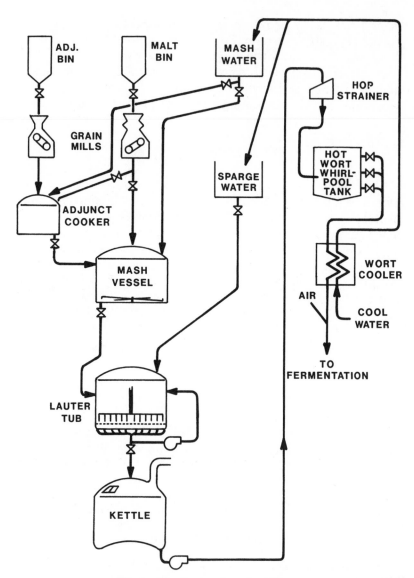

Fig. 3.   Brewhouse schematic diagram.

the same in breweries all over the world (Fig. 3). In the United States malt and grits are usually received by rail shipment, and the grains are conveyed to bins for storage. The first stage of the process is grinding, in which roller mills are used to crush the grains to permit penetration of water into the interior.

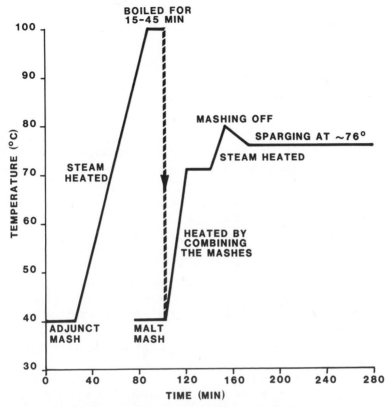

Fig. 4. Typical temperature program used for mashing. [From J. J. Dougherty (see Master Brewers Association of the Americas, 1977).]

All of the starchy adjunct and a portion of the malt (usually about 10%) are placed in the adjunct cooker together with some water. This mixture is stirred and heated rapidly to boiling. This sacrifices the malt enzymes in the cooker mash, but enough activity is derived from them during the temperature rise to liquefy the starch; otherwise, heating the starch would result in a thick, gelatinous mass that would be difficult to pump. Meanwhile, the remainder of the malt and water are placed in the mashing vessel. This slurry is stirred and heated according to a temperature program with a number of holds and rises (see Fig. 4). One of the temperature rises is produced by pumping the very hot cooker mash into the main mash. The temperature program is designed to maximize the activities of the malt enzymes of interest by maintaining temperatures at which they are very active but not denatured to a significant degree, until they have had enough time to produce the desired effect. The major reaction of the amylases is the conversion of starch into maltose, some other simple sugars, and limit dextrins:

$$\text{starch} + H_2O \xrightarrow[\beta\text{-amylase}]{\alpha\text{-amylase}} \text{maltose} + \text{glucose} + \text{limit dextrins},$$

$$\text{protein} + H_2O \xrightarrow{\text{proteases}} \text{amino acids} + \text{peptides}.$$

Malt protein is hydrolyzed by proteolytic enzymes into peptides and simple amino acids. The final temperature rise at the end of mashing is designed to inactivate the carbohydrase enzymes and is called *mashing off*.

The next operation is removal of the spent grains from the mixture by a filtration operation. Most often this is done in a round vessel, called a *lauter tub,* with a false bottom containing many fine slots. The mash is dropped by gravity or pumped into the lauter tub, and the liquid that drains through the false bottom is recycled to the top of the vessel for a time. The grains themselves become the filter bed, and the clear wort is then run into the kettle. After drainage is fairly complete, hot water is sprayed over the grains to obtain additional extract that has been retained in the bed; this operation is called *sparging.* The spent grains remaining in the lauter tub are removed and sold for animal feed.

The wort in the kettle is heated to boiling and usually boiled for about 90 min. Approximately 9–10% of the water is evaporated during this time, together with a number of low-boiling compounds. The heat of boiling inactivates all of the malt enzymes and kills virtually all microorganisms present. Hops are added toward the end of boiling, and the α-acids are dissolved in the wort; there they rearrange into the bitter iso-α-acids (see Scheme I). If leaf hops are used, then they are removed by pumping the wort through a hop strainer. This apparatus is not needed if hop extracts or pellets are used instead of leaf hops.

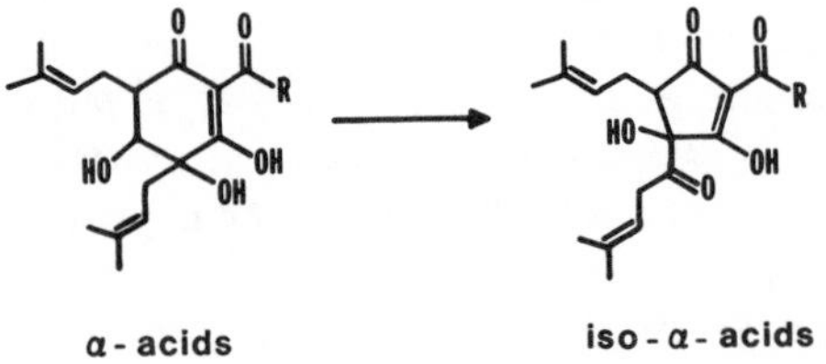

α-acids   iso-α-acids

Scheme I

The prolonged boiling causes a significant amount of precipitation of heat-coagulable material, called *trub*, which is very rich in protein. This is often separated from the wort in a whirlpool tank, which is a special sedimentation vessel. The wort enters this vessel tangentially, causing a swirling motion. As the sediment falls to the bottom, it is carried into the center of the tank by currents set in motion by frictional forces. A tight, pyramidal pile of solids forms in the center of the tank, and the wort is decanted from the vessel with minimal loss. The trub is removed and added to the spent grains for use as animal feed.

Next the wort is cooled to the desired starting temperature of fermentation. Usually this is done via a heat exchanger that transfers the heat to water that will be used for mashing or sparging. Refrigerant cooling may be used to complete the cooling process.

The wort is then aerated to obtain a uniform oxygen level. At this stage wort is a good nutrient medium for yeast, since it contains oxygen, many simple sugars (brewing yeasts cannot utilize sugars larger than three glucose units), and simple amino acids (brewing yeasts cannot utilize even small peptides). The wort is virtually free of microorganisms and contains a number of malt- and hop-derived flavor compounds.

B. Fermentation and Yeast Separation

As soon as yeast is added to the wort, fermentation begins. Yeast addition is most often done in-line as a fermenter is filled. Typically about 10–15 million yeast cells are added per milliliter of wort.

Fermenting vessels vary widely in geometry and size. Some older fermenters are round tanks made of wood, open at the top. Others are closed, cylindrical tanks oriented on their sides. Rectangular-loaf–shape tanks are also in use. Recently installed brewery fermenters are mainly quite large (often holding several times the volume of a single brew from the brewhouse) and of the cylindroconical type; this is an upright cylinder with an inverted cone bottom (tapering down to a point).

Fermenters must either be equipped with cooling bands or coils or be placed in a cold room to maintain temperature control. Methods of managing fermentation temperatures vary from using an isothermal profile (see Fig. 5), with or without a cooler stage at the end, to a several-step process employing different temperatures. When fermenters without cooling are placed in a fixed-temperature room, the

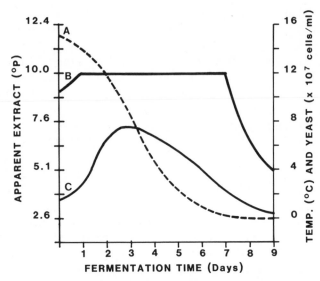

Fig. 5. Typical North American lager fermentation. Curve A, apparent extract; B, temperature; C, yeast in suspension. [From F. B. Knudsen (see Master Brewers Association of the Americas, 1977.] Degree Plato (°P) is the standard unit of measurement for apparent extract.

metabolic activity of the yeast warms the vessel; when all the nutrient has been consumed, the temperature falls. Ale fermentations, which are warmer, take as little as 3 days, while lager fermentations, which may be as cool as 6–10°C, often require 7–12 days.

The oxygen initially present in the wort is rapidly consumed by the yeast. At this point the fermentation becomes anaerobic, yeast energy metabolism changes, and production of ethanol begins. During the aerobic portion, the yeast has been exposed to enough oxygen to permit it to reproduce; it does this by budding. The energy required for this process is only available to the cell during aerobic metabolism, but it can be stored so that division may still occur 1–3 days after the oxygen has been depleted. Peak yeast counts during fermentation typically reach 50–80 million cells/ml, indicating that, on the average, each original yeast cell budded, and then the original and daughter cells both budded again.

The main reaction during fermentation is the conversion of maltose (the most prominent sugar) into ethanol and carbon dioxide (Scheme II). Many other sugars present, particularly glucose, sucrose, and maltotriose, are also consumed. Larger sugars remain untouched. The amino acids are taken up and used to synthesize

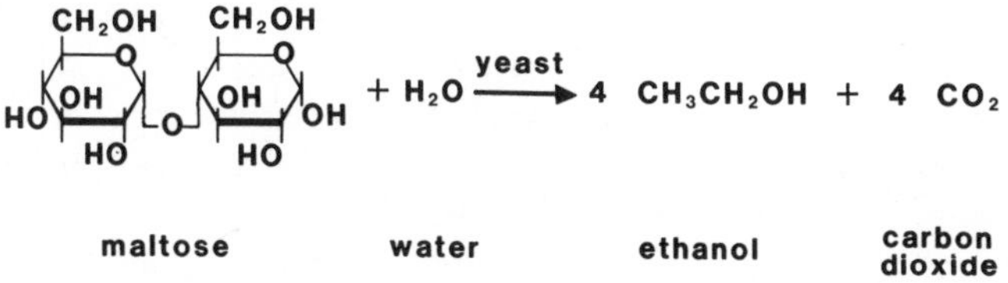

$$+ \; H_2O \xrightarrow{\text{yeast}} 4 \; CH_3CH_2OH \; + \; 4 \; CO_2$$

| maltose | water | ethanol | carbon dioxide |

Scheme II

new protein, both enzymes and structural protein, for new yeast cells. In addition to ethanol, a great many other compounds are formed by the yeast; these include ethyl esters, acetate esters, medium chain-length fatty acids, higher alcohols, and many others.

Although wort has a significant buffering capacity, enough organic acids and CO_2 are produced to lower the pH from 5.5 to 4.0–4.6. (This depends on the particular process and is quite characteristic.)

The CO_2 produced during fermentation is collected and processed to remove foam, water-soluble and water-insoluble volatile substances, and moisture. It is then compressed and liquefied for storage until needed for touch-up carbonation or exclusion of air during transfers and packaging.

Not all of the substances produced during fermentation are desirable. Vicinal diketone precursors are produced by yeast in the early part of fermentation. These compounds are spontaneously converted into vicinal diketones (VDKs), which cause a buttery flavor. This is considered to be a serious flavor defect. If yeast is still active when VDKs are formed, then it rapidly consumes these compounds. Two approaches to this problem can be taken. If the fermentation is kept warm near the end, then the conversion of the VDK precursors into VDKs takes place and the yeast remains active and removes the VDKs. If cooler fermentation regimes are used, then it is necessary to intentionally carry some yeast into the next stage to reduce the VDKs there.

One of the simplest ways to monitor the progress of fermentation is to measure the specific gravity of the fermenting wort. Before yeast is added, this depends only on the amount of dissolved solids present. The specific gravity is usually converted into the equivalent percentage of sucrose that would result in the same specific gravity, which for a wort is called the original gravity (OG). Once fermentation begins, the sugars are converted into ethanol, which has a density only 80% that of water, and CO_2, most of which is evolved as a gas. This causes a decrease in the observed specific gravity, which, expressed as percentage sucrose equivalent, is called the *apparent extract*. It is possible to obtain the alcohol content and the true amount of dissolved solids (real extract) remaining in the beer by making an independent measurement of some other property, such as refractive index, or by separating the alcohol from the wort by distillation before the specific gravity measurement.

Toward the end of fermentation, some of the yeast settles to the bottom of the fermenter. Depending on the type of vessel, the beer is either decanted from the yeast or the yeast is drawn off first, before the beer is removed. In some breweries the beer undergoes a secondary fermentation in a separate tank, where 10–15% kraeusen may be added. (Kraeusen, or curls, is a wort that has just started to ferment.) Once fermentation is complete, continuous centrifuges may be used to remove most of the yeast remaining in suspension. Yeast that is needed for subsequent brews is stored cold, but a significant quantity of surplus yeast is produced. Because yeast is very rich in protein, vitamins, and nucleic acids, it has a high nutritional value. In England and Australia most brewer's spent yeast is debittered, autolyzed, and sold for human consumption (as Marmite or Vegemite). In the United States some spent yeast is used in soups, and some is sold for animal feed. After yeast removal, the beer is cooled to near 0°C and pumped into a storage tank.

C. Storing, Blending, and Finishing

In the chill-storage, or lagering, step, beer is pumped into a tank and simply held at a temperature slightly above freezing for an extended period, usually between 10 and 30 days. For ales the storage step may be much shorter, at times as short as 5 days. In some cases a fining agent is added to the beer to speed clarification. Often this is isinglass, gelatin, or tannic acid. Fining agents accelerate the process of precipitation of haze material.

The major occurrence during lagering is sedimentation of substances that are difficult to filter and that if not removed, would tend to form haze in the finished package. Many of the particles in the beer at this stage are large enough to scatter light (colloidal size) but will not readily sediment (they are kept in colloidal suspension by Brownian motion). They will, however, grow larger due to aggregation and slowly sediment. The progress of clarification can be followed by monitoring light scattering, usually at 90° to the incident beam (nephelometry).

Some maturation of beer flavor takes place during storage. Reductions in the concentrations of acetaldehyde, hydrogen sulfide, and

VDKs occur. Some brewers intentionally carry over some yeast (on the order of 1 to 2 million cells/ml) into storage for its effect on flavor maturation.

At the end of storage, the beer is decanted from the storage tank, leaving the sediment behind on the tank bottom. The beer is then finely filtered, usually through diatomaceous earth.

Most brewers practice what is known as high-gravity brewing. This means that wort is prepared and fermented, and the beer is stored at higher original gravity than that at which it will be packaged. Water that has been deaerated and carbonated is blended with the high-gravity beer to reach the final package gravity. This approach increases the capacity of a brewery and offers some improvement in flavor stability as well. The blending operation may take place prior to lagering but most often is done just before final filtration.

The carbonation level of the beer is checked after filtration and, if necessary, adjusted, and the beer is metered and pumped into tanks which feed bottle-, can-, or keg-filling lines.

IV. Process-Control Measurements

Controlling the process just described requires measurement of temperatures, flow rates, liquid levels, solids weights, and pressures. Since the temperatures used in brewing are fairly modest, the materials handled are for the most part noncorrosive, and the flow rates employed are not particularly high, the sensors for these measurements are generally the same as those employed in many other processes. The needs for the product to avoid contamination with non–food-grade materials and to withstand caustic cleaning and hot-water sterilization are not particularly stringent requirements and are faced in the production of many other food products. Acidic cleaners are used for some purposes, as are surfactants and sanitizers (halogens and quaternary amines). The most difficult requirement to satisfy is avoidance of microbial infection after the kettle. Equipment must be designed so that it does not have any dead (nonswept) parts or portions that cannot be thoroughly cleaned.

A. Temperature

Temperature sensors used include liquid-filled or bimetallic thermometers, filled thermal systems, resistance bulb thermometers, thermocouples, and thermistors. Their use in brewing was reviewed by Soroko (1967) and Hahn (1977a). Precise temperature measurement and control are very important in mashing and also for uniform fermentations.

B. Flow Rate

Brewery flow-measurement devices include head-type flowmeters (orifice plates, venturi tubes, weirs, flumes, target meters, and rotameters), turbine meters, vortex shedding meters, and magnetic flowmeters. These were discussed by Soroko (1967) and by Hahn (1977a). Measurements of flow rates are needed to proportion yeast into wort and are particularly critical during the blending of high-gravity beer with water late in the process.

C. Liquid Level

Methods of detecting liquid levels in tanks include differential pressure transmitters, air bubblers, float-type switches, capacitance sensors, ultrasonic level detectors, and conductivity-level switches (Hahn, 1977a; Soroko, 1967). Tank-level measurements are useful in monitoring tank filling. Measuring the level can be complicated by the foam layer formed when wort or beer is pumped into a tank; ultrasound and capacitance sensors are suitable for use under these conditions.

D. Solids Weight

Accurate grain weights are needed to maintain brew-to-brew reproducibility. These are obtained using large scales employing mechanical levers, load cells (hydraulic, pneumatic, or strain gauge), or continuous weighing equipment (belt-type and solids-flow transmitter) (see Hahn, 1977a).

E. Pressure

Pressure measurement in breweries must cover the range from vacuums of 10 to 100 mm Hg in water-deaerating systems through pressures of 1000 to 2000 psig for steam generation. This may be done with bourdon-tube gauges or diaphragm pressure sensors of various kinds (Hahn, 1977a; Soroko, 1967).

F. Conductivity

An installation in which kettle heating was controlled to maintain a vigorous boil while not boiling over was described by Moyer (1969). Conductivity sensors were positioned to detect when the liquid reached a specific point on the vessel wall.

In the filtration operation, care is taken to exclude oxygen. Lines and filters are first filled with water to displace air, and beer is followed by water at the end of a filter run. Conductivity-measuring devices can be used to sense the changeover between beer and water and to minimize beer losses and out-of-specification beer (Küster and Schlosser, 1975; Schröder, 1977). The beer lines between bottling tanks and fillers are often long; the point at which a changeover between one product and another reaches a filler may be sensed with a conductivity probe.

V. On-Stream Measurements

The term "on stream" is used here somewhat loosely, because brewing is not a continuous process. Some instrument installations serve to automate the batch operation of a particular vessel. At other points, blending of various brews may be practiced, or brews may be immediately followed by others so that they appear as a continuous stream to a monitoring instrument in-line. The topics treated here are arranged in the order in which they are encountered in the process.

A. Wort Gravity

During the wort-filtration process, the solids content (OG) of the wort varies due to sparging. Conductivity analyzers (Neugebauer *et al.,* 1970) and refractometers (Matèrne, 1979) have been used to monitor the wort OG. The results can be used to guide the operation of lautering and sparging so that worts of uniform OG are produced.

Hahn (1977b), in a review of process-monitoring instruments, described three methods of sensing density or specific gravity. (i) Glass hydrometers can be mounted in special housings and equipped with position transmitters for remote indication. (ii) The weight of a horizontally pivoted, fixed-volume U tube can be sensed. (iii) Vibrating U-tube sensors produce an amplitude proportional to the mass of the liquid they contain. Yet another approach to this problem is the use of a gas bubble probe to measure density (Treiber and Höhn, 1976).

B. Dissolved Oxygen

Some breweries monitor wort oxygen content. Generally this is done shortly after wort aeration, but adequate opportunity for mixing should be allowed. This is not a difficult application in terms of sensitivity since the wort is close to saturation with air at this point (about 9–10 mg/liter). Care must be taken to remove bubbles before measurement since they can cause incorrect results (Moll *et al.*, 1977). It is possible to have more air present than the solubility limit and not reach saturation. The type of oxygen-sensing apparatus ordinarily used for this application is a polarographic probe covered with a gas-permeable membrane (Krebs, 1975). In some cases, brewers wish to obtain higher oxygen levels than are possible with air saturation. In this case oxygen rather than air can be used, and the measurement is more important for process uniformity.

Dissolved-oxygen measurements during and at the end of fermentation present quite different problems than they do in wort (Moll *et al.*, 1978). In beer the difficulty is in providing adequate sensitivity to O_2 levels as low as 10 μg/liter. Membrane polarographic probes are currently the instruments of choice, particularly when yeast is present in the sample (Hahn and Hill, 1980; Wisk *et al.*, 1981). Galvanic cells (Galloway *et al.*, 1967; Hunt *et al.*, 1968) have also been used, but they generate serious errors with yeast-containing samples.

Dissolved oxygen is a key parameter, both in blending water and in beer before and after blending. As in the fermentation case, membrane-covered polarographic probes work well. Galvanic cells have been used at this process point since little yeast is present, but they are difficult to calibrate and not as reliable as the polarographic analyzers (Wisk *et al.*, 1981).

C. Fermentation Monitoring

Several different approaches to monitoring the progress of fermentation have been used. These include monitoring apparent extract directly or making indirect observations of fermentation byproducts such as heat or CO_2. Devices for measuring apparent extract have been described and are useful for monitoring the rate of fermentation. In some breweries in which larger volume fermenters are used, such devices have been installed in each vessel. One such instrument is the Platometer, which employs two pressure sensors mounted a fixed distance apart (Møller, 1975). These are installed in the tank with one sensor directly above the other. The pressure difference measured by the two sensors is used to determine the specific gravity of the fermenting wort.

Ruocco *et al.* (1980) described the use of exothermic measurements to observe fermentation behavior. This method was used in cooling-equipped fermenters by periodically turning the cooling apparatus off and observing the rate of temperature increase. Cooling was then resumed until the time of the next observation.

Measuring CO_2 evolution as a means of monitoring fermentation was described by Moll *et al.* (1978). They applied a number of instruments to large cylindroconical fermenters.

D. Yeast Concentration

The continuous measurement of yeast concentration in a fermenter, using electronic particle counters and a specially designed interface, has been described (Moll *et al.*, 1978). The design of the interface is quite complicated because it is necessary to precisely dilute the yeast in an electrolyte buffer and to disperse yeast cell clumps before the instrumental observation.

E. Fermenter Gas Purity

While CO_2 can be collected from a fermenter for use, the air content has a significant effect on the amount of energy required for compression and liquefaction. It is therefore advantageous to begin collection only after a high percentage of the air initially present in the fermenter headspace has been swept out by the CO_2 produced. Carbon dioxide purity in fermenter gas can be gauged with a process katharometer or a Wheatstone bridge (Saeger, 1978). With a katharometer the heat capacitance of fermenter gas is compared with that of pure CO_2. Similarly, with a Wheatstone bridge the conductivity of the fermenter gas is compared with that of a reference gas.

F. Carbon Dioxide in Ambient Air

The air in fermenting cellars can reach significant CO_2 concentrations when fermenters leak or when many are opened for cleaning. This poses a safety hazard and falls under OSHA regulation. Instruments for measuring CO_2 based on the use of infrared light absorption in a folded-path gas cell are offered in portable form and for fixed installation in a cellar (Bedrossian, 1981). A small pump is used to draw air through the instrument cell.

G. Ozone

One treatment that may be applied to blending water is ozonation. This sterilizes the water and oxidizes (deodorizes) certain organic substances that may be present in the incoming water. A fine line must be held between too little ozone with insufficient biocidal action, and too much, with high energy usage and rapid deterioration of the piping system (even when it is made of stainless steel) (Egan *et al.*, 1979). On-stream ozone analyzers can be used for control of ozonation and are usually placed to monitor water at the end of treatment. This ensures adequate treatment in cases of high levels of oxidizable substances in the incoming water.

H. Alcohol, Original Gravity, and Calories

Several systems have been described that employ on-line measurement of two properties [such as specific gravity and refractive index (Lloyd, 1979) or the measurement of refractive index at two different temperatures (Panter, 1978)] and the use of simultaneous equations to calculate alcohol, real extract, OG, and even calories. Any one of these results can in principle be used for feedback control of blending. This is advantageous since different parameters are more critical in different products (i.e., alcohol in 3.2 beers and calories in light beers). Unfortunately, the precision required for specific-gravity and refractive-index measurements places severe demands on the instruments.

I. Turbidity

Instruments for measuring turbidity (haze) in filtered beer usually employ right-angle scattering (nephelometry), which is sensitive to colloidal-size particles (Hahn, 1977b). In-line nephelometers are offered by Sigrist (Zürich, Switzerland) and Jacoby Tarbox (Yonkers, New York). Narrow-angle scattering instruments, such as that made by Monitor Technology (Redwood City, California) give a greater response to larger particles and have been popular for monitoring filtration performance with diatomaceous-earth filters (Simms, 1972). It is common to use these to automate filter operation by sensing breakthrough of the filter. Brewers generally wish to measure turbidity at a point where the beer is quite cold (much of the haze dissolves on warming), often colder than the air temperature where the instrument is mounted. This can lead to problems of condensation on the optics of turbidity-measuring instruments.

J. Carbon Dioxide in Beer

Shortly before packaging, the CO_2 content of the beer must be checked and adjusted, since this affects filler performance and the foam and flavor of the finished product. The measurement can be made manually but has been automated using several different approaches [see the review by Hahn (1977b)].

Some instruments perform grab sampling, remove CO_2 from the beer via agitation, and sense the amount by means of pressure and volume relationships (deBrune *et al.*, 1974; Lloyd, 1979). Other instruments separate the gases from the beer with a gas-permeable membrane, sweep the CO_2 into a gas cell with nitrogen, and measure its quantity by infrared light absorbance (Gamache, 1968; Houben and Turner, 1974). An infrared analyzer with a multiple internal-reflectance cell in contact with beer has been used to sense CO_2 without prior separation (Bedrossian, 1981).

A recently developed analyzer employs a selective gas-permeable membrane to pass CO_2 and establish equilibrium between the CO_2 dissolved in the beer and that in the gaseous state on the other side of the membrane (Kesson, 1982). A pressure sensor connected to a microcomputer quantitates the amount of CO_2 present.

Packaged, automated systems are available in which analysis of CO_2 content is performed in-line, and the results are used to control injection of CO_2 into the beer (Hahn, 1977b; Rohner and Tompkins, 1970; Wieland, 1979).

VI. Chromatographic Methods

Chromatographic methods used in brewing cover a wide spectrum of concentrations and a broad array of general-purpose and highly specific detectors. The procedures range from direct injection of the sample (good potential for on-stream analysis) to complicated extraction and enrichment (not promising for continuous monitoring). Many trace-component determinations are highly dependent on the sample matrix (Siebert, 1983). Gas chromatographic (GC) methods that have been applied to brewing were reviewed by Chen and Morrison (1983). A similar treatment of high-pressure liquid chromatographic (HPLC) procedures was written by Cieslak (1983). Because the main concern here is quantitative results, only GC and HPLC methods will be discussed. Methods that apply only to raw materials, such as examination of hop oils, or to beer after production, such as determination of flavor-staling compounds, will not be described.

A. Carbohydrates

Perhaps the most important products of wort production are the fermentable sugars. These can be analyzed by HPLC after a very simple sample-cleanup procedure: Mixed-resin ion exchangers are simply added to wort. After a short contact time, some of the wort is drawn off and filtered through a fine-pore–membrane filter. The samples are then injected directly into an HPLC equipped with a refractive-index detector. The separating mechanisms used vary. Metal-loaded resin-based ion-exchange columns (Aminex HPX-87, Aminex HPX-42, both from Bio-Rad, Richmond, California) elute higher oligomers followed by simple sugars using water as the eluant (Dadic and Belleau, 1982). The Aminex HPX-87 column does not separate sugars larger than maltotriose. Higher oligomers elute together as a "dextrin" peak. The Aminex HPX-42 column resolves oligosaccharides as high as degree-of-polymerization (DP) 8. Dextropak (Waters Associates, Milford, Massachusetts) with water as eluant resolves saccharides from DP 1 through DP 13 in about 40 min. Acetonitrile–water mobile phases were used with both μBondapak–carbohydrate (Waters Associates) and a silica column treated with tetraethylenepentamine and produced similar results (Buckee and Long, 1982). Both procedures showed good resolution of monosaccharides through trisaccharides in 15 min, but higher oligomers were retained by the columns.

The same carbohydrate procedures described above can be applied to syrups, fermenting worts, and beers. Since carbohydrates higher than DP 3 are not fermentable by brewing yeasts, all of the above procedures are suitable for normal worts and beers. The main difference between light and regular beers is in the oligosaccharide content. Only those carbohydrate methods that give results for both low- and moderate-DP compounds are useful for light beers.

B. Amino Acids

Measuring the pattern of wort amino acids is useful when deliberate alterations of wort composition are made, because yeast takes up different amino acids at quite different rates. The levels of particular amino acids affect biochemical pathways including VDK formation. Dedicated amino acid analyzers (Enari *et al.*, 1970) or HPLC-based methods for determining amino acids can be used. Since beer is a physiological fluid (containing proteins and peptides), it is more difficult to analyze than protein hydrolysates are. For routine control purposes, methods that determine total amino acids appear to be satisfactory. Amino acid analysis does not therefore appear to be needed as an on-line method.

C. Hop Bitter Resins

The bitter resins from hops and worts can be extracted and determined via HPLC using UV-absorbance detection. Both the nonbitter precursors of the bitter substances (α-acids) and the bitter iso-α-acids can be determined in this way. Separations of these substances have been carried out by normal partition, ion exchange, adsorption, and reverse-phase mechanisms. Current favorites employ bonded reverse-phase columns and either ionic suppression or ion pairing (Verzele and Dewaele, 1981; Knudson and Siebert, 1983).

D. Vicinal Diketones

Vicinal diketones can be assayed in fermenting wort and beer via headspace methods employing electron-capture detection (Gales, 1976; Harrison *et al.*, 1965). These methods take advantage of the fact that very few volatile compounds in beer capture electrons. Automated headspace-analyzer GCs have been used for this purpose and give more-precise results than manual methods because of the difficulties in making reproducible manual headspace injections. The precursors of the VDKs can also be determined by comparing results of analyses in which the sample is analyzed as is with those in which the same sample is first heated to convert the VDK precursors into VDKs.

E. Beer Aroma Volatiles

The major beer aroma volatiles produced along with ethanol can be determined by GC with a flame-ionization detector, via headspace GC (Hoff and Herwig, 1976), with a purge and trap sampler (Chen, 1983), or by extractive methods (Stenroos *et al.*, 1976). These volatiles usually include higher alcohols (*n*-propanol, isobutanol, isoamyl alcohols, phenylethanol), esters (ethyl acetate, isobutyl acetate, isoamyl acetate, phenylethyl acetate, ethyl hexanoate, ethyl octanoate), and fatty acids (mainly hexanoic, octanoic, and decanoic). Both packed and capillary columns have been used for this analysis. All of these methods can be applied to fermenting worts and beers in process. An installation has been described in which a process GC was installed to monitor these compounds in a large-scale fermenter (Moll *et al.*, 1978).

F. Sulfur Compounds

Malt contributes dimethyl sulfide (DMS) to beer, and a certain amount of DMS (typically about 50 μg/liter) and DMS flavor is normal in lagers (less so in ales, which are traditionally made with malts that are more highly kilned). Malt contains both free DMS and its precursor, S-methyl methionine (SMM). Any free DMS is lost at kettle boiling, but some SMM persists and is converted into DMS during the warm stage before the wort is cooled. Malts and worts can be assayed for both DMS and SMM, through its forced conversion into DMS (White, 1977). The determination is made by headspace GC with a flame-photometric detector (FPD) operated in the sulfur-specific mode (Hysert *et al.*, 1979). Calibration is very important because of the nonlinear response of this detector.

The sulfur compounds present in finished beer include H_2S and thiols produced by yeast during fermentation. The latter compounds are very difficult to measure due to their low concentrations (typically 5 μg/liter for H_2S and 1–2 μg/liter for thiols). Reliable quantitation requires the use of GC with sulfur-specific detection, usually FPD (as for DMS), and additionally requires that metal contact be scrupulously avoided (Takahashi *et al.*, 1978). Teflon columns with powdered Teflon packings and particular liquid phases are needed, since even glass contains sufficient metal ions to interfere with detection of the tiny amounts of these compounds present.

G. Phenolic Compounds

Phenolic compounds, primarily monomeric phenols, were determined in an ethyl acetate extract of beer via HPLC with electrochemical detection (Roston and Kissinger, 1981). The separation was performed on a bonded reverse-phase column with an eluant of propanol–methanol–aqueous-ammonium-acetate buffer.

H. Nitrosamines

Nitrosodimethylamine can be detected in beer down to about 1 µg/liter via GC using one of several different methods that are highly specific for the nitroso group. Most such methods employ a distillation or extraction step. The most common detector used is the thermal energy analyzer, which employs the thermoluminescent reaction of nitrogen oxides derived from the nitroso functional group with ozone (American Society of Brewing Chemists, 1981). The Hall conductivity detector with a gold catalyst has also been used, as has GC–mass spectometry with selected-ion monitoring.

I. Anions

Certain anions in water, wort, and beer can be conveniently detected through the use of ion chromatography; this is HPLC employing an ion-exchange mechanism to separate substances including chloride, sulfate, phosphate, acetate, and nitrate (Knudson and Siebert, 1984). Different detectors can be used to monitor the separation; in this case indirect refractive-index detection worked best. The only sample pretreatment required was membrane filtration to remove particles.

J. Ethanol

Ethanol is one of the inherently simple compounds to quantitate in beer due to its high concentration. Since it is reported to the nearest 0.01%, determination must be quite precise. A rapid (2 min) GC method that employs direct injection of beer after degassing and addition of an internal standard (Jamieson, 1979) was granted official method status by the American Society of Brewing Chemists. A flame-ionization detector is used. Glass inserts before the column are recommended to trap nonvolatile materials.

Cieslak and Herwig (1982) described an HPLC method in which beer is directly injected on an Aminex HPX-85 column operated at 85°C. The eluant used was very dilute sulfuric acid in water. A refractive-index detector provided excellent quantitative results.

K. *Proteins*

Lusk *et al.* (1983) described the separation of beer proteins with HPLC using size-exclusion columns. A UV-absorbance detector was operated at 280 nm.

VII. Overview

It is of interest to consider how the four main properties of beer are related to the measurements described. These are listed from the simplest and best understood to the most complex.

A. *Color*

Beer color is primarily derived from malt color, with some dilution from the use of adjuncts. The precise compounds responsible for color are not known, but color intensity is easily measured photometrically. The best control is a malt-color specification and a uniform process.

B. *Haze*

Haze is less clearly defined than color. Beer haze contains protein, polyphenol, and carbohydrate. Of the polyphenols, dimeric and trimeric compounds are the most haze active. None of these (especially the polysaccharides) are well defined in any of the procedures noted. The HPLC analysis of phenolic compounds gives results for monomeric rather than dimeric or trimeric phenols. Some prominent oligosaccharides can be quantified by HPLC, but the polysaccharides are larger in molecular size and smaller in quantity than those determined. Recent work with molecular-size–based separations of proteins has been interesting and may help understanding of haze.

As most of the proteins and polyphenols come from malt, the use of adjuncts dilutes them. Precipitation of haze material during storage leaves less to produce haze in the package. Selective removal of protein fractions or phenolics with adsorbents also reduces haze potential.

C. Foam

Foam is less well understood than haze and is difficult to measure. It is related to the CO_2 content and the levels of iso-α-acids and fairly high molecular-weight proteins, particularly those which are rich in basic amino acids and possibly associated with carbohydrates. Lipids exhibit foam-negative behavior. Of the substances noted, CO_2 and iso-α-acids are readily quantitated. Proteins can be separated according to size, but no information regarding composition is available from this assay.

Foam is improved with more malt protein (in contrast to haze), higher hopping rates (which make the beer more bitter), and higher CO_2 (which also has some effect on flavor).

D. Flavor

Knowledge of flavor is even less complete. It is related to ethanol, esters, higher alcohols, short and medium chain fatty acids, H_2S, thiols, DMS, hop bitter resins, and other substances not described here (malt- and hop-oil-derived compounds), many of which are not well understood. Still others have yet to be discovered.

Flavor depends on malt and hops, with the strong influence of yeast in modifying some of the malt and hop flavors and in forming many other compounds. The spectrum of yeast-derived compounds is strongly dependent on the yeast strain used, on subtle factors of wort composition, and on the fermentation temperature.

E. Summary

Except for CO_2, all of the other known factors that influence haze, foam, and flavor must be measured chromatographically, if at all.

The difficulty in measuring these substances on-line, the sheer complexity of the phenomena, and the gaps in brewing knowledge account for the present difficulty in producing quality beer with continuous process stages.

References

American Society of Brewing Chemists (1981). *J. Am. Soc. Brew. Chem.* **39**, 99–106.
Bedrossian, J., Jr. (1981). *Master Brew. Assoc. Am. Tech. Q.* **17**, 87–91.
Buckee, G. K., and Long, D. E. (1982). *J. Am. Soc. Brew. Chem.* **40**, 137–140.
Chen, E. C.-H. (1983). *J. Am. Soc. Brew. Chem.* **41**, 28–31.
Chen, E. C.-H., and Morrison, N. M. (1983). *J. Am. Soc. Brew. Chem.* **41**, 14–18.
Cieslak, M. E. (1983). *J. Am. Soc. Brew. Chem.* **41**, 10–13.
Cieslak, M. E., and Herwig, W. C. (1982). *J. Am. Soc. Brew. Chem.* **40**, 43–46.
Dadic, M., and Belleau, G. (1982). *J. Am. Soc. Brew. Chem.* **40**, 141–146.
deBrune, P., Cremer, J., Dorrenboom, J. J., and Witte, J. H. M. (1974). *Brauwelt* **114**, 1430–1432.
Egan, L. H., Taylor, K. R., and Hahn, C. W. (1979). *Master Brew. Assoc. Am. Tech. Q.* **16**, 164–166.
Enari, T-M., Linko, M., Loisa, M., and Makinen, V. (1970). *Master Brew. Assoc. Am. Tech. Q.* **7**, 237–240.
Gales, P. W. (1976). *J. Am. Soc. Brew. Chem.* **34**, 123–127.
Galloway, H. M., Raabe, E. A., and Bates, W. (1967). *Am. Soc. Brew. Chem. Proc.*, 79–83.
Gamache, L. D. (1968). *Am. Soc. Brew. Chem. Proc.*, 120–124.
Hahn, C. W. (1977a). *Master Brew. Assoc. Am. Tech. Q.* **14**, 59–69.
Hahn, C. W. (1977b). *Master Brew. Assoc. Am. Tech. Q.* **14**, 87–93.
Hahn, C. W., and Hill, J. C. (1980). *J. Am. Soc. Brew. Chem.* **38**, 53–60.
Harrison, G. A. F., Byrne, W. J., and Collins, E. (1965). *J. Inst. Brew.* **71**, 336–341.
Hoff, J. T., and Herwig, W. C. (1976). *J. Am. Soc. Brew. Chem.* **34**, 1–3.
Houben, W. S., and Turner, G. S. (1974). *Brew. Guardian* **103**(7), 23, 25.
Hunt, W., Espadas, O., and Lee, S. L. (1968). *Master Brew. Assoc. Am. Tech. Q.* **5**, 167–170.
Hysert, D. W., Morrison, N. M., and Jamieson, A. M. (1979). *J. Am. Soc. Brew. Chem.* **37**, 30–34.
Jamieson, A. M. (1979). *J. Am. Soc. Brew. Chem.* **37**, 151–152.
Kesson, J. (1982). *Brew. Guardian* **111**(10), 18–20.
Knudson, E. J., and Siebert, K. J. (1983). *J. Am. Soc. Brew. Chem.* **41**, 51–56.
Knudson, E. J., and Siebert, K. J. (1984). *J. Am. Soc. Brew. Chem.* **42**, 65–70.
Krebs, W. M. (1975). *Master Brew. Assoc. Am. Tech. Q.* **12**, 176–185.
Küster, J., and Schlosser, H. (1975). *Brauwelt* **115**, 75–81.
Lloyd, M. (1979). *Master Brew. Assoc. Am. Tech. Q.* **16**, 182–185.
Lusk, L. T., Cronan, C. L., Chicoye, E., and Goldstein, H. (1983). *J. Am. Soc. Brew, Chem.* **41**, 31–35.
Master Brewers Association of the Americas (1977). "The Practical Brewer", (H. Broderick, ed.) 2nd ed. Madison, Wisconsin.
Matérne, H. (1979). *Brauwissenschaft* **32**, 29–33.
Meilgaard, M. C. (1982). *J. Agric. Food Chem.* **30**, 1009–1017.

Meilgaard, M. C., Reid, D. S., and Wyborski, K. A. (1982). *J. Am. Soc. Brew. Chem.* **40**, 119–128.

Moll, M., d'Hardemare, Ch., and Midoux, N. (1977). *Master Brew. Assoc. Am. Tech. Q.* **14**, 194–196.

Moll, M., Duteurtre, B., Scion, G., and Lehuede, J.-M. (1978). *Master Brew. Assoc. Am. Tech. Q.* **15**, 26–30.

Møller, N. C. (1975). *Master Brew. Assoc. Am. Tech. Q.* **12**, 41–45.

Moyer, G. G. (1969). *Master Brew. Assoc. Am. Tech. Q.* **6**, 203–207.

Neugebauer, K., Weber, H., and Kupprion, R. (1970). *Brauwelt* **110**, 701–704.

Panter, J. (1978). *Brew. Distilling Int.* **8**, 34.

Rohner, R. L., and Tompkins, J. R. (1970). *Am. Soc. Brew. Chem. Proc.,* 111–117.

Roston, D. A., and Kissinger, P. T. (1981). *Anal. Chem.* **53**, 1695–1699.

Ruocco, J. J., Coe, R. W., and Hahn, C. W. (1980). *Master Brew. Assoc. Am. Tech. Q.* **17**, 69–76.

Saegar, R. W. (1978). *Master Brew. Assoc. Am. Tech. Q.* **15**, 222–225.

Schröder, J. (1977). *Brygmesteren* **10**, 235–236.

Siebert, K. J. (1983). *J. Am. Soc. Brew. Chem.* **41**, 4–9.

Simms, J. R. (1972). *Master Brew. Assoc. Am. Tech. Q.* **9**, 25–30.

Soroko, O. (1967). *Master Brew. Assoc. Am. Tech. Q.* **4**, 2–8.

Stenroos, L. E., Siebert, K. J., and Meilgaard, M. C. (1976). *J. Am. Soc. Brew. Chem.* **34**, 4–13.

Takahashi, T., Nakajima, S., Konishi, I., Miedaner, H., and Narziss, L. (1978). *Brauwissenschaft* **31** (1), 1–4.

Treiber, K., and Höhn, K. (1976). *Brauwissenschaft* **29**, 221–226.

Verzele, M., and Dewaele, C. (1981). *J. Am. Soc. Brew. Chem.* **39**, 67–69.

White, F. H. (1977). *Brew. Dig.* **52** (5), 38–44, 46–48, 50.

Wieland, D. (1979). *Master Brew. Assoc. Am. Tech. Q.* **16**, 120–123.

Wisk, T. J., Weiner, J. T., and Siebert, K. J. (1981). *J. Am. Soc. Brew. Chem.* **39**, 147–153.

18

Reduction of H₂S and HCN in Coke Oven Gas

DAN P. MANKA

Pittsburgh, Pennsylvania

I. Introduction

Large volumes of coke oven gas (COG) are generated by the carbonization of coal for the production of blast-furnace coke. At one plant, 80 million ft³ of COG are produced daily as a result of coking 7500 tons of coal.

This gas is treated for removal of tar, ammonia, naphthalene, and aromatics before it is used as a fuel in many of the steel-plant furnaces.

A. EPA Ruling on SO₂ Pollution

Regulations state that the sulfur content of COG must be reduced to 50 grains/100 ft³ (800 ppm) so that SO_2 production during combustion is drastically reduced. This is true for any sulfur-bearing gas, including natural gas.

To comply with these regulations, facilities were installed to remove sulfur compounds and hydrogen cyanide (HCN) from COG. A continuous sampling and analysis system was developed and installed to continuously analyze sulfur compounds and HCN. This system and various methods of extracting these compounds will be described in this chapter. [Also, see Manka (1975).]

B. Sulfur Compounds in COG

The major sulfur constituents of a COG after condensation of tar and removal of ammonia, naphthalene, and aromatics are given in Table I.

C. H₂S Concentration in COG

1. CALCULATION OF SULFUR CONTENT OF COG

The sulfur content of the COG can be calculated by a formula based on the total-sulfur analysis of the coal used to charge the ovens (Lowry, 1945).

A rough guide for predicting the approximate H_2S content in COG is as follows:

(i) The H_2S content in grains per 100 cubic feet equals 365 times the percentage of sulfur in the coal.†

(ii) The organic sulfur in the gas in grains per 100 cubic feet equals 18 times the sulfur content of the coal.

2. WET CHEMICAL ANALYSIS OF SULFUR AND CYANIDE IN COG

The procedure for the wet chemical analysis of hydrogen sulfide in COG is well documented, as is the method for HCN analysis (Manka 1979).

---

† Grains of H_2S per 100 cubic feet times 15.4 equals approximate parts per million.

TABLE I

CYANIDE AND SULFUR CONSTITUENTS
OF AROMATIC-FREE COKE OVEN GAS

Compound	Volume (PPM)
HCN (hydrogen cyanide)	800
COS (carbonyl sulfide)	50
H_2S (hydrogen sulfide)	7000
SO_2 (sulfur dioxide)	50
CS_2 (carbon disulfide)	50
CH_3SH (methyl mercaptan)	3
C_2H_5SH (ethyl mercaptan)	3

II. Continuous On-Line Analysis of H₂S and HCN

A. Preliminary Considerations

Several important factors must be considered in the design of a sampling system for the analysis of sulfur and cyanide compounds in COG. The gas is washed in the plant system directly with water to remove naphthalene and cool the gas (see Fig. 1). (In the final step, the gas is extracted with a petroleum oil to remove light oil; therefore, it is saturated with water at 40°C.) Some naphthalene is carried through in the gas and can crystallize and settle out to plug small lines and valves. The gas also contains some tar oils and compounds, such as dienes, that polymerize to form rubbery substances at higher temperatures and in the presence of acid catalyzers. These factors must be considered in the design of a sampling system.

B. Sampling

The best sampling point is a straight stretch of gas line located downstream from several elbows; these elbows thoroughly mix the gas to produce sample homogeneity.

One sampling system that has been used successfully for many years is described as follows. The COG flows from the plant duct through a Teflon probe to the approximate center of a drop-out tank and then through either of two lines of filters and pumps. The gas is pumped to a box where a small volume is diverted through a final filter to the analyzers. The major portion of the gas passes through the rotometer and is returned to the plant duct located downstream from the probe.

D. P. MANKA

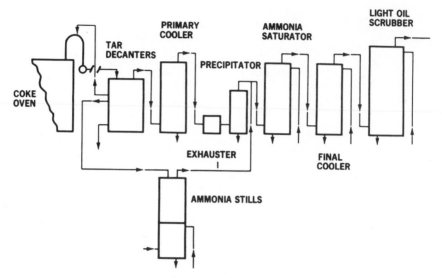

Fig. 1. Flow diagram of COG [from Manka (1982)].

The probe is a Teflon tube of $1\frac{1}{4}$-in. and $\frac{3}{4}$-in. Teflon has proven to be a good material because less tar and solids adhere to it than to a stainless steel tube. The high gas velocity helps to prevent the accumulation of solids. The tube extends across the full width of the 36-in. plant duct so that a representative sample is supplied to the analyzers. The tube is inserted into a 2-in. diameter stainless steel pipe with one-half of the pipe cut along the length of the probe, except at the top and bottom, so that the six $\frac{3}{8}$-in. probe openings are exposed to the gas. This arrangement securely holds the probe in position so that the openings are on the same course as the gas flow.

The first hole on each end of the probe is located $3\frac{1}{2}$ in. from the plant-duct wall. One end of the probe is closed by a $\frac{1}{4}$-in.–thick Teflon disk. The other end , located outside of the duct, is connected by means of Swagelok fittings to 1-in. stainless steel tubing, which is for the most part used throughout the sampling system. A source of steam may be connected to the probe exit to clean the six probe openings should they become clogged.

The sample gas flows to the center of a knock-out drum that is designed to collect entrained water, oil, and tar in case a plant operation malfunction causes large amounts of water and oil to be carried in the gas stream. (In normal operation, only a few drops of water collect in the bottom of the drum.) This drum is made from 12-in.–diameter pipe, which is large enough to accept excess quantities of entrained oil and water.

Gas-line temperature from the probe to the knock-out drum is automatically maintained at 100°F or higher if the gas contains considerable amounts of naphthalene.

Gas flows from the top of the knock-out drum through either of two lines to filters located in a heated compartment. (One line-and-filter assembly is a spare; this arrangement permits replacing filter elements without interrupting operation.) The filter is inverted so that the dirty gas enters through an opening drilled into the drain, which is normally installed on the bottom of most filter cannisters. Experience has taught that all 90° bends on lines carrying dirty gas should be replaced with a lazy bend; the 90° bends are easily plugged by dirty gas. When using the lazy bend, dirty gas flows directly into the filter instead of the usual inlet on the cannister head, which has a 90° bend. The filters clean the gas efficiently, removing the solids that are larger than 1 μm. With this clean gas, there is no problem with 90° bends in the remainder of the system.

The clean gas flows to either of two metal bellows pumps that have welded, leak-tight steel bellows. The pump section is located in a heated compartment to prevent condensation in the gas, while the motors are outside of the compartment. The flow rate is 120 ft³/hr at a pressure of 15 psig.

The gas flows from the pump to a small heated compartment where a small portion of the gas, at a flow rate of about 100 cm³/min, is diverted through a glass filter to a cyclone located in the heated compartment below the analyzers. Approximately 15 cm³/min flow to the rotometers and then to the two sampling valves in the chromatographs. The excess gas from the cyclone is vented. The major portion of the gas in the 1-in. pipe line is returned to the plant gas duct located downstream from the sampling probe.

Temperature of all gas lines and filters beyond the knock-out drum is automatically maintained at 66°C to prevent condensation of moisture.

It is imperative that a clean and representative gas sample be injected into the chromatograph to obtain the proper analysis. This sampling system serves the purpose exceptionally well. It has been used for several years with excellent analytical results.

C. Gas Chromatography

Gas chromatography is a versatile tool for the analysis of organic and inorganic gases. It is particularly selective and sensitive to sulfur compounds when the chromatograph is equipped with a flame photometric detector (FPD). The use of this detector has major applications in the analysis of sulfur-containing air pollutants in the parts-per-billion and parts-per-million concentration ranges.

The major characteristic of the FPD is its nonlinear response to sulfur compounds. When calibrating a detector, the response may be plotted as a function of the sample concentration; a logarithmic plot of this data gives a straight line. Response functions can be expressed by the exponent n in the relation

$$\text{response} = (\text{concentration})^n. \qquad (1)$$

This exponent is also dependent on the compound of interest. (For H_2S, $n = 2$.) When n is determined by tests on samples of known concentrations, that value may be used for the quantitative analysis of unknown samples. However, as the concentrations in the sample increase, n changes due to the peculiar behavior of the detector. Therefore, analyses should be done within the calibration range where the logarithmic plot of the detector response is a straight line. Various analyses may be confirmed by the wet chemical method.

A series of tests on the FPD with samples of sour COG indicated that the sample should be diluted considerably to reduce H_2S concentrations to levels in the normal FPD-response range. When the gas is diluted approximately 200-fold with nitrogen in a dilution flask, the H_2S concentration is within the range of constant n. However, diluting the sample causes the concentrations of other sulfur compounds to become too low for reliable analysis in a plant stream, particularly when the sample size is controlled by the H_2S concentration. Also, analysis of HCN concentration requires that the chromatograph be equipped with a thermal conductivity detector (TCD). Therefore, the column design must permit analysis of low concentrations of sulfur compounds as well as high concentrations of H_2S and HCN.

D. Column Technology

An improvement in column technology, developed by the author in collaboration with Honeywell (Houston, Texas), resulted in a unique and successful approach to these analyses. The initial problem is the elimination of compounds that would interfere with the complete separation of the specified gaseous components. This means that the sulfur compounds are rapidly and completely separated from oils, naphthalene, and water, and that these unwanted components are backflushed to atmosphere. The sulfur compounds are then separated chromatographically; sufficient time must be allowed between peaks so that the high concentration of H_2S can be eliminated by venting it to atmosphere before FPD analysis. By this method only the low-concentration sulfur compounds reach the detector; a large sample of gas can be used to analyze these compounds. Since the concentrations of methyl and ethyl mercaptans are less than 1 ppm in COG, only COS, CS_2, and SO_2 require analysis.

High concentrations of H_2S and HCN are analyzed by a TCD. Oils and water are separated from the sample and backflushed to vent. The H_2S, HCN, and other compounds have to be separated completely to that they can be analyzed by the TCD.

The schematic for separation and analysis of the sulfur compounds (but not high-concentration H_2S) in the stream analyzer is shown in Fig. 2. In the flame photometric chromatograph, a 600-μl sample of coke-oven gas is injected by sample valve into the $\frac{1}{8} \times$ 12-in. Teflon column containing 100/120-mesh Teflon coated with 5% SF-96, a methyl silicone stationary phase. The sulfur compounds pass unresolved through this column and the backflush valve to the second column, leaving behind oils, naphthalene, tar, and water, which are then backflushed to vent. The sulfur compounds are separated into COS, H_2S, CS_2, and SO_2 on the $\frac{1}{8} \times$ 11-in. Teflon column filled with a treated silica gel. When the high concentration of H_2S reaches the selector valve, the valve is switched to vent the H_2S to atmosphere. There is sufficient time between the peaks of COS, H_2S, and CS_2 to permit complete elimination of H_2S to vent so that it does not enter the FPD. Therefore, only COS, CS_2, and SO_2 are analyzed in this chromatograph.

The column and valves occupy one compartment, which is maintained at 63°C. Nitrogen is the carrier gas at a flow rate of 16 cm³/min. All columns are made of Teflon, and the valves have a Vespel body and a Teflon diaphragm. The entire analysis is completed in 13 min.

The flow schematic for the separation of H_2S and HCN is shown in Fig. 3. The thermal conductivity unit is in a separate compartment with its own valves and auxiliary equipment. It operates at the same time as the FPD so that both analyses are completed in 13 min. A 1000-μl sample of COG is injected by the sample valve onto the precolumn, which consists of a $\frac{1}{16} \times$ 3-in. stainless steel tube filled with 100/120-mesh Porapak Q plus a $\frac{1}{16} \times$ 12-in. stainless steel tube filled with 100/120-mesh Chromasorb P coated with 15% Carbowax

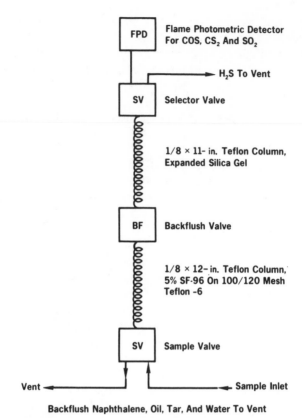

Fig. 2. Column schematic for analysis of sulfur compounds in low concentration [from Manka (1979)].

6000. The HCN and unresolved sulfur compounds (except CS_2) elute through the backflush valve onto the second column, leaving behind the oils, naphthalene, tar, and water plus the CS_2 held by the short Porapak Q column. On a signal from the analyzer control, the backflush valves are switched, and one source of helium backflushes these oils and water to vent while a second source of helium flows to the second column where HCN and the sulfur compounds are resolved. This resolution is done on two columns in series, the $\frac{1}{16} \times$ 3-in. stainless steel tubing containing 50/80-mesh Porapak T plus the $\frac{1}{16} \times$ 48-in. stainless steel tubing containing 100/120-mesh Porapak Q. All the components flow through the thermal conductivity detector, but only the responses to H_2S and HCN are recorded. This procedure allows the detection of H_2S concentrations up to 20,000 ppm

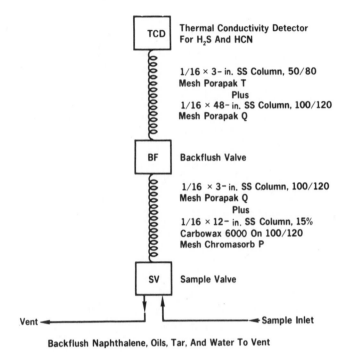

Fig. 3. Column schematic for analysis of HCN and high concentrations of H₂S [from Manka (1979)].

and HCN up to 4000 ppm. The short Porapak T column is necessary to completely separate propylene and HCN. Micropacked columns are used because of their high efficiencies and rapid component separation.

The detector is a thermistor thermal conductivity cell with a volume of approximately 0.05 μl. It is a semidiffusion cell that is relatively insensitive to flow rate. The response is suitable for analysis times of less than 1 sec. Both the columns and the detector are located in one compartment maintained at 60°C. Helium is the carrier gas at a flow rate of 13 cm³/min. The bodies of the sample and backflush valves are stainless steel, and the diaphragms are Teflon.

The analyzers are calibrated with standard gases contained in small aluminum cylinders. These cylinders are enclosed and maintained at a temperature of 40°C.

A better calibration curve to determine the actual exponential factor in Eq. (1) for each component may be generated by plotting a log–log curve of peak area versus concentration. The slope of each curve is the experimental factor for each component for the particular FPD and conditions under which the calibration curve was generated. With a microprocessor-based programmer, these exponential factors can be automatically calculated for use in the linearization program.

Although the theoretical value of the exponent for H_2S is 2.0, several authors have found variations in these values for this and other sulfur compounds. [Burnett *et al.* (1977) have found such variations.]

The concentration of H_2S depends on the temperature of carbonization, the reaction of gases with the hot, white coke, and the reaction of gases with the walls of the ovens. There also could be catalytic effects related to minerals in the coal.

E. Control Systems

The control section is an all–solid-state unit that utilizes plug-in modules. The use of plug-in modules simplifies equipment troubleshooting. The section consists of the following modules: monitor, sequencer, amplifier, valve and dual valves, square root, and memory for each of the components.

The control system could be improved by the addition of a microprocessor programmer to determine the calibration exponent and to use the derived exponent in calculating sulfur concentration.

The monitor module is essentially a maintenance unit that is used to monitor various parameters of the equipment. The sequence module contains the operating-mode selector switch and the precision crystal-controlled digital clock on which all system timing is based. The amplifier module provides final amplification of the preamp output from the detector switch. The calibration circuitry in the module produces a reference voltage that is used in setting the memory zero and span. The component module contains the programming switches and control circuitry for the component. It is adjusted for "gate" start time, the duration of each component peak, and the gain to be employed during the peak elution. It also contains an auto zero, which automatically adjusts the baseline prior to elution of the component. The valve modules contain the programming switches and circuitry for backflush, sample, and vent valves.

The concentrations of components are determined by peak height or area. Using the peak-height method, the maximum height of the peak is compared with the voltage response to the same gas in the standard. Using the peak-area method, the area under the peak is compared with that of the same component in the standard gas. An Anadex digital printer records the concentration of each component in parts per million. A bar-graph recorder allows observation of the various separations and aids in calibration. The day of the year and time of day are also printed out with each analysis.

The analyzer and control systems are located in an air-conditioned trailer where the temperature is maintained at 22°C. The trailer is located near the source of the sample.

F. Results

The sampling system has operated continuously for several years at a flow rate of 120 ft³/hr at a pressure of 15 psig. There has been no plugging in the lines or filter. Ambient temperatures as low as $-18°C$ have had no effect on the system.

The analyses of 140 automatically injected samples of COS, H_2S, SO_2, CS_2, and HCN over a period of 30 hr showed a reproducibility of $\pm 2\%$.

Approximately once a week the gas flowing to the analyzer is analyzed for H_2S by wet chemical methods for the purpose of verifying the chromatographic results. Analyses by the two methods have been in close agreement.

A separate sampling system has been installed to analyze the low concentrations of sulfur gases in the sweet gas exiting the desulfurization plant by the FPD. The old sampling system is used to bring sour gas to the analyzer to determine the high concentrations of H_2S and HCN using the thermal conductivtiy detector. A stream selector switches to bring a sample of the gas from the plant system immediately after desulfurization. This gas is analyzed for the low H_2S and other sulfur compounds using the FPD. Provision has also been made to analyze the sample with the TCD.

The differences in analyses of the two gas streams show the extraction efficiency of H_2S, HCN, and other sulfur compounds by the desulfurization process.

V. Extraction of H₂S and HCN from Gas

The process of extracting H_2S and HCN is too extensive to be described completely in this chapter. The reader is advised to contact chemical companies that have installed desulfurization processes in coke plants; these include Applied Technology Corporation (Houston, Texas), Dravo Corporation (Pittsburgh, Pennsylvania), and Carl Still Corporation (c/o Dravo Corporation). Only a short description will be given here so that the reader can compare the methods used with coke-oven gas and natural gas.

A. Aqueous Ethanolamine Solution

The monoethanolamine (MEA) method is called the *Sulfiban process*. It consists of four principle parts:

(i) absorbing acid gases in a COG contactor by lean aqueous ethanolamine solution;

(ii) regenerating the rich MEA solution in a still column;

(iii) heating and cooling the MEA solution to control the temperature of process vessels; and

(iv) filtering and reclaiming the solution to maintain solution purity.

The purpose of the Sulfiban process is to remove sulfur compounds from the raw COG after the light oil scrubber. Sulfur is carried in the gas principally as H_2S, which is a serious environmental pollutant. Removal of H_2S and the other sulfur compounds, CS_2 and COS, is accomplished by absorption into the MEA solution, which has a great affinity for sulfur compounds. The solution and the COG are brought into contact with each other in a vertical contactor column.

The solution enters at the top and flows down through a bed of packing. The gas enters the contactor at the bottom and flows up through voids in the packing. The packing is designed to provide high surface area for gas–liquid contact. After completing its flow through the column, the gas exits the vessel at the top and flows into the plant gas mains as a sweet, environmentally acceptable fuel gas. After completing its flow down through the column, the solution collects at the bottom of the vessel as a H_2S-rich solution; it is then pumped to the regenerating apparatus where it is heated to remove H_2S and filtered before it is recycled back to the contactor column as fresh solution. (Regeneration is accomplished by reversing the contactor column process. The H_2S-rich solution enters at the top of the still column, and as it falls down over the packing in the vertical vessel, a counterflow of steam heats it and strips out the gas from the solution. This lean solution is then cooled and recycled to the contactor.) Only small quantities of solution must be replaced provided that demisters are installed in the gas line exiting the contactor to prevent carry-over of MEA solution in the gas.

Steam and acid gases exit at the top of the still column, and the steam is condensed. The water and gas are separated in the reflux accumulator below the still; the concentrated acid gas is taken off for further processing.

The reclaimer functions in the cycle to maintain solution purity by removing any products that may be formed by the reaction of the solution with components in the COG.

B. Aqueous Ammonia Solution

The COG exiting the electrostatic precipitors, which remove most of the fine tar components, may be desulfurized with an aqueous ammonia solution. The gas is cooled from 55 to 30°C by water circulating in the precooler. (The water temperature is maintained by an evaporative cooler.) To dissolve condensed naphthalene, tar-containing flushing liquor from the circulating ammonia liquor is mixed with the recycled water.

From the precooler section, the COG goes to the H₂S scrubber section. This is a tower with special low–pressure-drop packing and distribution and draw-off trays. The H₂S removal is primarily accomplished by countercurrently contacting the gas with ammonia solution under conditions that optimize selective H₂S absorption while minimizing CO₂ absorption. Final cleanup is accomplished by dilute caustic-soda solution, which is used to spring fixed ammonia in the ammonia still.

From the H₂S scrubber, the COG goes to the plant ammonia scrubber, where the ammonia is removed with sulfuric acid, in the usual manner, to make ammonium sulfate.

The sulfide-rich solution from the bottom of the H₂S scrubber section is pumped to deactifier columns. In these columns the acid gases (and some ammonia) are stripped using steam. The ammonia that comes over with the acid gases is removed in an ammonia-wash column using cooled flushing liquor as the scrubbing medium. The H₂S-rich gas is pumped to the sulfuric acid plant or to the Claus sulfur plant.

C. Thylox Process

The extraction of H₂S from COG with a thioarsenate solution is called the Thylox process.

The raw gas is washed in a grid-packed column with arsenical liquor. The acid-bearing liquor from the bottom of the column is discharged into a reservoir tank and then passed through a heating system and into the regeneration column, where it comes in contact with compressed air. Elemental sulfur produced in the reaction of air with the acid-bearing liquor at 110°C is separated from the liquor at the top of the regenerator by frothing. The regenerated liquor is cooled and flows back to the absorption column.

Basic reactions of the Thylox process are

(i) absorption

$$Na_4As_2S_5O_2 + H_2S \longrightarrow Na_4As_2S_6O + H_2O$$

and

(ii) regeneration

$$Na_4As_2S_6O + O \longrightarrow Na_4As_2S_5O_2 + S.$$

The original thioarsenate molecule is thus regenerated and the overall effect is

$$H_2S + O \longrightarrow H_2O + S.$$

Some sulfur in the regenerating column is oxidized to thiosulfate. This creates demand for alkali (Na_2CO_3 or NH_3). Thiosulfate is a fixed salt in the Thylox liquor. Constant bleeding of the liquor from the system is necessary to control the concentration of thiosulfate.

Very little HCN is removed by Thylox. HCN is removed before the H_2S tower by scrubbing the gas with ammonium polysulfide to form thiocyanate:

$$NH_3 + (NH_4)_2S_n + HCN \longrightarrow NH_4SCN + (NH_4)_2S_{n-1}.$$

References

Burnett, C. H., Adams, D. F., and Farwell, S. O. (1977). *J. Chromatogr. Sci.* **15**, 230–232.

Lowry, H. H. (1945). "Chemistry of Coal Utilization," vol. II, p. 948. Wiley, New York.

Manka, D. P. (1975). *Instrum. Technol.* **22**(2), 45–49.

Manka, D. P. (1979). *In* "Analytical Methods for Coal and Coal Products," vol. III (C. Karr, ed.), pp. 3–33. Academic Press, New York.

Manka, D. P. (1982). *In* "Automated Stream Analyses for Process Control," vol.1 (D. P. Manka, ed.), pp. 273–282. Academic Press, New York.

19

Conversion of H₂S to Sulfur or Sulfuric Acid

DAN P. MANKA

Pittsburgh, Pennsylvania

I. Introduction

Hydrogen sulfide (H_2S) and hydrogen cyanide (HCN) are recovered from sour coke-oven gas by monoethanolamine (MEA), ammonia liquor, or thioarsenate liquor, as described in Chapter 18. The process of HCN decomposition and H_2S conversion to elemental sulfur or sulfuric acid are too lengthy to be described completely in this book; only a brief overview will be presented.

II. HCN Decomposition

The concentrated HCN in the concentrated acid gas is corrosive to the sulfur recovery plant, promotes plugging, and produces a poor-quality sulfur product. Therefore, immediately prior to the Claus burner, the HCN in the acid gas is catalytically decomposed in a series of reactions involving hydrolysis oxidation into ammonia, carbon disulfide, carbonyl sulfide, and carbon dioxide:

(i) $HCN + 2H_2S + \frac{1}{2}O_2 \longrightarrow CS_2 + H_2O + NH_3,$
(ii) $HCN + H_2O \longrightarrow NH_3 + CO,$
(iii) $CS_2 + H_2O \longrightarrow COS + H_2S,$
(iv) $COS + H_2O \longrightarrow CO_2 + H_2S.$

The inlet acid gas is heated to 121–176°C by the HCN preheater. The preheated acid gas is blended with oxygen and steam and continues to the HCN reactor containing bauxite catalyst. The oxygen reacts according to reaction (i), resulting in a temperature rise. At higher temperatures (260–315°C) the hydrolysis reactions [(ii), (iii), and (iv)] begin. There is a slower rise in temperature related to these reactions.

The HCN is primarily converted to NH_3, with small changes in H_2S, COS, and CS_2 concentrations. The HCN-free acid gas then continues to the acid-gas burners.

III.   Conversion of H_2S

A.   *To Sulfuric Acid*

The following describes the operation of a single-absorption concentrated sulfuric acid recovery system. This system includes water removal from the process gases immediately prior to the SO_2 converter. The off-gas coming from the desulfurization plant consists of H_2S, CO_2, HCN, NH_3, and water vapor.

The sulfuric acid (H_2SO_4) wet-gas contact-type production system consists of the following:

(i)   HCN decomposition,
(ii)   generation of SO_2 gas by combustion of the H_2S off-gas,
(iii)   catalytic oxidation of SO_2 to sulfur trioxide (SO_3), and
(iv)   absorption of SO_3 gas into circulating H_2SO_4 where it is hydrolyzed to H_2SO_4 by water in the acid.

The process begins by combustion of the off-gas in the burner with heated air providing the oxygen:

$$2H_2S + 3O_2 \longrightarrow 2SO_2 + 2H_2O + \text{heat}.$$

Combustion destroys any HCN, NH_3, and hydrocarbons remaining in the off-gas.

The hot gases are cooled and then passed to a contact-type cooler where the hot gases are quenched by direct contact with weak H_2SO_4. Weak acid is sprayed into the cooler chamber. The wet gas flows through a demister chamber where any weak acid mist is removed. The gas is then dried in a packed tower with 93% sulfuric acid.

The conversion of SO_2 to SO_3 by catalytic oxidation is represented by

$$2SO_2 + O_2 \longrightarrow 2\,SO_3 + \text{heat}$$

The final step of the process is the absorption of SO_3 gas from the converter into 98% acid, which contacts the gas in a packed column. The absorbed SO_3 is hydrolyzed by the water present in the 98% acid:

$$SO_3 + H_2O \longrightarrow H_2SO_4 + \text{heat.}$$

The sulfur recovery system treating acid gas from the HCN system is a Claus method. The reactions occurring in the Claus process are

$$H_2S + \frac{3}{2}\,O_2 \longrightarrow SO_2 + H_2O,$$

$$2H_2S + SO_2 \longrightarrow \frac{3}{n}\,S_n + 2H_2O,$$

where n defines the possible molecular forms of sulfur vapor.

Sufficient air to burn approximately one-third of the H_2S (to SO_2) and all the other combustibles in the feed gas (hydrocarbons, NH_3, etc.) is added to the sulfur plant burner and mixed with all of the acid gas; this mixture is then burned. The combustion products from the burner flow through an oven where enough time is allowed for complete burning of hydrocarbons and ammonia.

The hot gas from the oven then passes through a waste-heat boiler, giving up a substantial amount of its heat in the generation of steam. In a separate condenser, hot gas from the boiler is then cooled further to condense sulfur vapor that was formed in the oven. Liquid sulfur from this condenser drains continuously to the sulfur storage tank.

Cool gas from the condenser is reheated to catalytic-reaction temperature with steam. The hot gas enters the first stage of a multistage reactor where a certain amount of H_2S and SO_2 gases are reduced to elemental sulfur. The first reactor stage also converts most of the organic sulfur compounds, COS and CS_2, to H_2S and elemental sulfur. The sulfur produced is then separated from the gas stream in a condenser, the balance of the gas stream flows to the next reaction stage, and the condensed elemental sulfur stream passes to the sulfur storage tank. The successive cycles of reheating, reacting, and condensing remove sulfur until the desired degree of conversion has been achieved.

To initiate the catalytic reaction in the reactor vessels, effluent gas from each condenser must be reheated to the temperature at which the reaction can proceed at the desired rate. Various methods of reheating are used.

Any remaining liquid sulfur entrained in the tail gas will be removed in the tail-gas separator, which is equipped with a heated mist eliminator.

Tail gas is finally burned in an incinator to oxidize any remaining H_2S to SO_2 before it is released to the atmosphere.

C. *To Sulfur in the Thylox Process*

The conversion of H_2S to sulfur by the use of arsenical solution or thylox liquor has been adequately described in Chapter 18.

20

Analysis of Chemicals Derived from Coal Carbonization

DAN P. MANKA

Pittsburgh, Pennsylvania

I. Introduction

The iron and steel industry accounts for up to 20% of total industrial consumption in the United States. Coal is a major source of this energy. Converting coal into coke serves the steel industry in three ways:

(i) Coke is essential as a source of heat and as a reactant in the blast furnace to convert iron ore into pig iron.

(ii) One-third of the gas generated during the coking of coal serves as fuel for underfiring the coke ovens.

(iii) The remaining two-thirds of the coke oven gas (COG) is used in the reheating furnaces in the steel plant.

187

II. Chemicals from COG

Valuable chemicals are recovered from the COG. Benzene is used in the production of nylon, and toluene and xylene are used as solvents and in the making of various chemicals. Phenols recovered from tar and coke gas are used in making thermosetting phenolic plastics. Ammonia is used as fertilizer in agriculture. Coumarone-indene recovered from the high-boiling solvent is used in the resin industry. Naphthalene recovered from tar is used to make phthalic anhydride. Several of these compounds are in sufficiently high concentrations to make recovery in pure form economically feasible. Analyses of most of these compounds were given previously (Manka, 1979).

III. Coal Carbonization

In the production of coke, coal is heated in ovens for several hours in the absence of air at temperatures above 1000°C to remove volatile compounds. The coal must be washed with water to remove many impurities, and it is charged to the ovens in wet or dry form.

A coke oven is contructed from high-temperature resistant bricks, with dimensions as follows: 15–18 in. wide, 15 ft high, and 41 ft long. The greatest width of 18 in. is on the "coke side," where the coke is pushed out of the oven into coke cars. (Subsequently, the red-hot coke is cooled directly with water.) The narrow width on the other end of the oven, generally called the "pusher side," is where a large ram from a pusher car is guided into the oven to push out the coke. Between the outer walls of the ovens are openings extending the full height and length of the oven; these are the flues. The COG is burned in these flues to supply heat to the charged coal. The wall on one side of the oven is heated for a predetermined time; the other wall is then heated for the same period of time. In this manner, heat is uniformly supplied to the coal.

There are generally 59 ovens to a battery. All of the volatile liquid flowing from a single battery is collected in a separate collecting main; the liquid from all batteries is mixed before entering the tar–liquor separating tanks. In this discussion we shall consider a coke plant consisting of 4 batteries with 59 ovens each and 1 battery with 79 ovens, for a total of 315 ovens.

Not every type of coal can be charged to the ovens. The concentrations of ash and sulfur must be considered to make good, strong

TABLE I

Typical Coals Charged to Coke Ovens

Source	Percentage in mixture	Volatile matter (%)	Fixed carbon (%)	Ash (%)	S(%)	H_2O (%)	Characteristics
Low volatile W. Va.	12	16.3	75.7	8.0	0.7	6.5	12.5% expansion 13 psig pressure
Mid volatile Pa.	15	24.7	68.4	6.9	0.89	7.5	8.0% expansion 4 psig pressure
High volatile Pa.	46	36.4	56.7	6.9	1.35	7.0	26% contraction 0.9 psig pressure before contraction
High volatile W. Va.	27	33.6	61.4	5.0	0.64	7.4	15.5% contraction 2 psig pressure before contraction
Mixture	—	31.4	62.9	5.7	1.03	6.3	10% contraction 1.7 psig pressure before contraction

coke. If there is too much sulfur in the coke, then it will find its way into the steel. Some coals expand when they are heated, while others contract. Some coals have a high content of volatile matter, while others have a low concentration. Therefore, blends of various coals are coked in small experimental ovens to determine the expansion or contraction of the coal and the physical characteristics of the final coke. If the coal expands too much, then it will be detrimental to the walls of the ovens; it may cause the coke to stick so that it cannot be pushed out of the ovens. Too much contraction of the coal forms an undesirable, dense coke. The optimum coal mixture is determined by the quality of the final coke, which must meet certain hardness, volatile-matter, and ash- and sulfur-content criteria to be suitable for use in the blast furnace.

Table I shows the analysis of a typical coal mixture charged to the ovens. The lower the sulfur content, the lower the H_2S concentration in the COG. To meet the pollution standards set by the EPA, H_2S must be removed from the gas before it is burned. Note that the contracting coals expand somewhat during the coking process before finally contracting. The expansion and contraction of the various coals was determined on the individual coals charged to the small experimental oven.

The oven is charged through three openings located at its top, which is filled to a height of 12 ft with 16.2 tons of coal. A leveling bar on the pusher machine levels the coal in the oven. As noted previously, all doors and openings on the oven are closed so that the coal is heated in the absence of air; otherwise the coal would simply burn and not form coke. Oxygen analysis of the gas exiting the oven during the 16–17 hr coking cycle cycle is 0.1% or less. A total of 6950 tons of coal is charged per day, forming 4600 tons of coke and 73,300,000 ft^3 of COG.

IV. COG Flow Diagram

The flow diagram of gas is practically the same in all coke plants. The flow diagram in Fig. 1 shows that COG rises from the coke oven, through the standpipe to the gooseneck, where it is contacted with flushing liquor (ammonia liquor). Tar and moisture are condensed. Ammonium chloride and a portion of the ammonia, fixed gases, HCN, and H$_2$S are dissolved by the liquor. The gas, liquor, and tar enter the gas-collecting main, which is connected to all ovens of a battery. The gas, liquor, and tar are separated in tar decanters. The separated tar flows to tar storage. A portion of the liquor is pumped to the gooseneck on the top of each oven. The remainder of the liquor is pumped to the ammonia-liquor still, where it is contacted with live steam to drive off free ammonia, fixed gases, HCN,

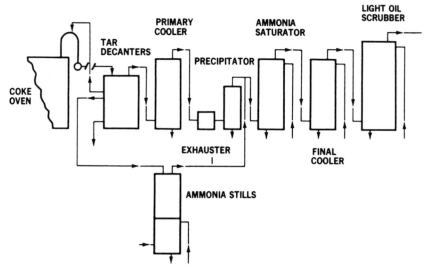

Fig. 1.   Flow diagram of COG [from Manka (1982)].

and H_2S. The liquor flows from the free still to the fixed still, where it is contacted with lime or sodium hydroxide to liberate free ammonia from ammonium chloride. Live steam flows up through the fixed and free stills, and the ammonia, fixed gases, HCN, and H_2S are added to the main COG stream ahead of the ammonia saturator.

The COG, separated from liquor and tar, is cooled indirectly with water in the primary coolers. The fine tar that separates from the gas is pumped to the tar storage tank. The cooled gas is pumped by exhausters to the electrostatic precipitators where additional fine tar is precipitated and pumped to the tar storage tanks. The gas is contacted with a dilute solution of sulfuric acid in the ammonia saturator to remove free ammonia. The ammonium sulfate–laden acid flows to the ammonia crystallizer (not shown in the figure) where crystals of ammonium sulfate are separated, and the remaining sulfuric acid is pumped back to the ammonia saturator.

The ammonia-free gas flows to the final coolers where it is further cooled by direct water contact. The water, with condensed naphthalene, flows from the cooler through tar, which absorbs the naphthalene. The water is cooled and recirculated into the final cooler.

The cooled gas enters the wash-oil, or benzole, scrubbers, where it contacts pumped-in wash oil (a petroleum oil). The aliphatic and aromatic compounds are extracted from the gas by the wash oil. The principal components are benzene, toluene, xylenes, indene, and solvent, also known collectively as light oil. The benzolized wash oil is pumped to the wash-oil still (not shown in the figure) where live steam strips out the light-oil compounds. The debenzolized wash oil is cooled and returned to the wash-oil scrubber. In some plants the light oil is further processed and fractionated into benzene, toluene, xylenes, and a high-boiling solvent fraction. Naphthalene is also present in the light oil. The gas from the wash-oil scrubbers is pumped by boosters to a gas holder, which tends to equalize the pressure. Booster pumps distribute one-third of the gas for underfiring the coke ovens and two-thirds to the steel plant where it is used as a fuel in many furnaces.

V. Determining the Coking-Cycle End Point by Gas Analysis

The length of a coking cycle is determined by experience and the range of flue temperature. Most coke plants also have experimental coke ovens to determine coking cycles for various coal mixes at

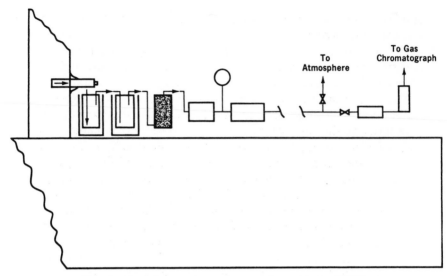

Fig. 2.   Sampling gas from coke oven [from Manka (1979)].

different flue temperatures. Another method used to determine the
end of this cycle is the analysis of the gas flowing from the oven into
the standpipe.

The gas sample from the standpipe must be cooled, separated
from tar and water, and filtered before it is analyzed for H_2, O_2, N_2,
CH_4, CO, CO_2, and illuminants. The sampling train depicted in
Fig. 2 has been used successfully for the continuous pumping of gas
to a sample bottle or to a gas chromatograph (GC).

Normally, there is an opening on the side of the standpipe where
live steam is admitted to remove accumulated tar in the pipe. This
opening is ideally located for gas sampling. The probe, extending to
approximately the center of the standpipe, is a $\frac{3}{4}$- or 1-in. heavy-wall
stainless steel pipe. The end of this pipe in the gas stream is sealed,
and the other end has a plug for removing accumulated tar. A $\frac{1}{4} \times 3$-
in. slot is cut along the length of the pipe, extending from the closed
end in the gas stream. When the probe is inserted into the standpipe,
it is important that the slot faces downstream; this position decreases the amount of tar pumped in with the sample. The probe is
fitted with a plug and rapidly pushed through the opening in the
standpipe until the plug seals the opening.

The two separators cool the sample and separate tar and water.
Both separators are kept submerged in water. The remaining piping
is $\frac{1}{4}$-in. stainless steel tubing with Swaglock fittings. Although most

of the tar is condensed in the separators, the gas is drawn through glass wool in a glass tube to remove the lighter tar oil. Tygon tubing is used for the metal-to-glass connection. The gas flows through a final filter that removes submicro particles.

The gas is drawn from the standpipe by a peristaltic pump; a vacuum gauge ahead of the pump is valuable in detecting line or separator plugging. When vacuum reaches 20 in., it is probable that tar is accumulating in the probe. A rod inserted through the plug opening of the probe will generally be sufficient to clear any obstruction. Gas passes through the valve downstream from the pump and is dried in a tube containing Drierite. The flow to the sample loop of the chromatograph is maintained at 50 cm³/min. Excess sample gas is vented to atmosphere.

VI. Gas Chromatographic Analysis of Detarred COG

The use of a portable GC located in a sheltered area near the sampling system is the best method of following the course of the coking cycle. The curves in Fig. 3 are based on analytical results

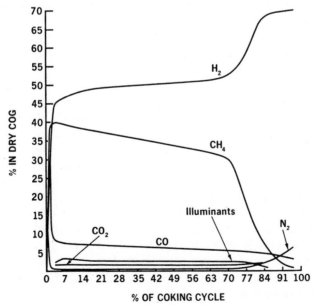

Fig. 3. Composition of COG during coking cycle [from Manka (1979)].

TABLE II

COMPOSITION OF DETARRED COG

Constituent	Volume percentage
H_2	55.0
CH_4	29.0
CO	5.5
Aromatics	3.0
CO_2 and S compounds	2.6
O_2	0.9
N_2	4.0

obtained by a GC with a thermal conductivity detector (TCD). The separation column is a $\frac{1}{8}$-in. × 10-ft stainless steel tube containing #5A molecular sieve, and the reference column is a $\frac{1}{8}$ × 67-in. stainless steel tube containing Porapak Q. The operating conditions are 12 cm³/min argon flow to each column, 100-mA cell current, and 40°C column and cell temperature. The chromatograph is standardized with a gas containing 50% H_2, 30% CH_4, 10% CO, 5% CO_2 and 5% N_2. The standard gas is admitted into the sample line upstream from the Drierite, and the flow is maintained at the same rate as for the COG. The sample-loop capacity is 0.5cm³ but may require adjustment based on the sensitivity of the detector. Argon is used as the carrier gas.

Concentrations of CO_2 and illuminants, such as ethylene, can be obtained on the Porapak column. The chromatograph must also be equipped with a sample valve for this column and a polarity switch so that the recorded peaks are positive. However, CO_2 and illuminant concentrations are not necessary to determine the end point; therefore, these can be ignored or only periodically determined by the Orsat method. The analytical results depicted in Fig. 3 are those for coking dry coal for 14 hr at a flue temperature of 1260°C. Concentration of H_2 approaches a maximum of over 70%, and that of CH_4 approaches a minimum of less than 1% near the end of the coking cycle. Generally, coking is continued for an additional 30 min after these values are reached, and then the coke is pushed from the oven.

Similar results are reported for the coking of wet coals. However, the coking cycle is 18–19 hr, because a large amount of water must be evaporated from the wet coal charged to the ovens.

The major constituents of the 73,300,000 ft³ of COG after tar condensation are shown in Table II.

VII. Major Chemicals Recovered from COG

A. Tar

The volume of tar produced in this particular coke plant is 58,900 gal/day. If the coke plant does not have its own tar refining plant, the tar is sold to a tar distilling company.

The tar is distilled into a fraction containing light oil, which is a mixture of benzene, toluene, and xylene; a second fraction containing phenol and its homologues; a third fraction containing coumarone–indene; a fourth fraction containing naphthalene; and a final fraction containing anthracene, carbazole, and phenanthrene. The light oil is fractionated into its main constituents, benzene, toluene, and xylene. The phenol fraction is extracted with sodium hydroxide to recover the phenols as sodium phenolate. The phenolate is neutralized with CO_2 to liberate free phenols, which are then fractionated into phenols, cresols, and xylenols. The fraction containing coumarone–indene is sold to plastics manufacturers. The naphthalene fraction is recovered by cooling the oil; naphthalene is refined by fractionation and used to prepare phthalic anhydride. If there is a market for anthracene, carbazole, and phenanthrene, then this oil fraction is refined by various methods, including fractionation.

B. Light Oil

In the coke plant discussed in this chapter, the wash oil extracts 22,300 gal of aromatics daily from the COG. The oil contains 63.0% benzene, 13.0% toluene, 5.6% xylenes, 5.1% crude solvent, and some naphthalene. The petroleum oil containing light oil is stripped of its light-oil content with steam, cooled, and returned to the light-oil scrubbers. The light-oil is fractionated in a forerunnings column to strip out paraffins, CS_2, and cyclopentadiene. Figure 4 shows a chromatogram of the light-oil effluent from the forerunnings fractionating column. The column is a $\frac{1}{8}$-in. × 13-ft stainless steel tube filled with 80/100 Chromasorb PAW coated with 10% (cyancethoxy) propane at a helium flow rate of 14 cm³/min. The concentration of CS_2, and especially cyclopentadiene, which polymerizes to a solid, must be controlled in the light oil. Therefore, these two constituents are analyzed chromatographically several times a day.

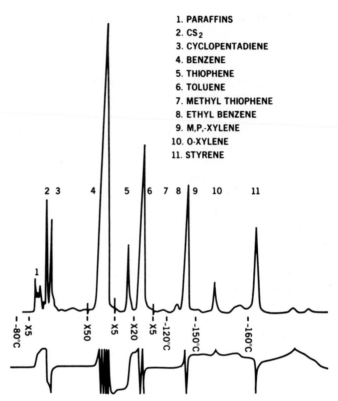

1. PARAFFINS
2. CS$_2$
3. CYCLOPENTADIENE
4. BENZENE
5. THIOPHENE
6. TOLUENE
7. METHYL THIOPHENE
8. ETHYL BENZENE
9. M,P,-XYLENE
10. O-XYLENE
11. STYRENE

Fig. 4. Chromatogram of light oil [from Manka (1979)].

C. Thiophene-Free Benzene

In the late 1950s and early 1960s there was a large demand for thiophene- and paraffin-free benzene for the production of nylon. Some benzene was freed by freezing it from thiophene. This was a slow, low-capacity process that did not remove all the paraffin. A catalytic hydrogenation process was developed in which the thiophenes reacted with H_2 to form H_2S, and the paraffins were extracted with ethylene glycol. The light oil was first distilled to a lower dry point of approximately 125°C so that the concentration of benzene was very high and the concentration of higher-boiling compounds, such as styrene and naphthalene, was very low. The chromatogram shown in Fig. 5 illustrates the analysis of this benzene-rich fraction during the many hydrogenation tests. The column is a ¼-in. × 15-ft stainless steel tube filled with 60/80 mesh Chromasorb coated with 10% glycerine carbonate. The column is activated by

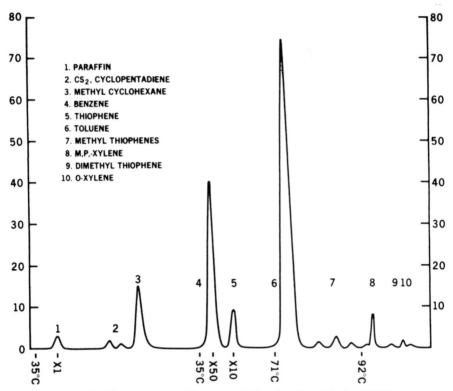

Fig. 5. Chromatogram of benzene-rich fraction [from Manka (1979)].

heating to 160°C at a helium flow rate of 30 cm³/min. The column is held at this temperature for about 5 hr. The gases eluting from the column during activation should not be passed through the thermal conductivity cell. Not many coatings were available in the early years of GC; glycerine carbonate was chosen because thiophene and methyl thiophene were completely separated from benzene, toluene, and xylenes. Furthermore, thiophene concentrations in the parts-per-million range could be easily detected. In the early years of chromatography, it was not unusual to use Rinso washing powder as a packing for some separations.

References

Manka, D. P. (1979). *In* "Analysis of Coal and Coal Products," (C. Karr, ed.), vol. III, pp. 3–27. Academic Press, New York.

Manka, D. P. (1982). *In* "Automated Stream Analysis for Process Control," vol. I (D. P. Manka, ed.), pp. 273–282. Academic Press, New York.

Index